To cow pasture
South
To Ned's Lick
Fields, Cataloochee,
Purchase, and points
North and East
Road to Waynesville
13 miles
3 to Soco Gap Road
log above
Barn to be
Garden
cow barn
CABBAGE

Mountain Fever

J. Baylor Roberts–Cataloochee Ranch Collection

Mr. Tom and Miss Judy at Fie Top in the early 1940s.

Mountain Fever

Tom Alexander

Edited by Tom Alexander, Jr.,
and Jane Alexander

Bright Mountain Books, Inc.
Fairview, North Carolina

Dedication

In the name of their father, who would certainly have agreed, the children of Tom Alexander would like to dedicate this book to their mother, Judith Barksdale Alexander.

✣ ✣ ✣

Permissions and credits for photographs are listed on page 162

Printed in the United States of America

ISBN: 0-914875-26-4

Library of Congress Cataloging-in-Publication Data

Alexander, Tom, 1900–1972.
Mountain fever / Tom Alexander ; edited by Tom Alexander, Jr. and Jane Alexander.
p. cm.
ISBN 0-914875-26-4 (hardcover)
1. Alexander, Tom, 1900–1972. 2. Foresters—United States—Biography. 3. Natural history—Appalachian Mountains, Southern. 4. Appalachian Mountains, Southern—Social life and customs.
I. Alexander, Tom, 1930– . II. Alexander, Jane, 1935– . III. Title.
SD129.A A3 1995
975.6'9042'092—dc20
[B] 94-23589
CIP

Contents

George Masa—Courtesy of M. R. Gooch and George Ellison

The view from Andrews Bald in the late 1920s, looking toward Bryson City. The man at left is Horace Kephart, noted chronicler of the southern Appalachians.

From Walker Bald, looking toward Dellwood and Lake Junaluska, with Miss Judy on Target in the foreground. Taken in about 1940, this picture shows the patchwork of clearings—subsistence farms and pastures—that once dappled the Appalachians. Nowadays, except in the valley bottoms, these clearings have mostly reverted to forestland, victims of changing agricultural economics and alternative sources of income. While incomes may be higher now, what has been lost is self-sufficiency and the independent attitudes that it engendered.

Siddell Studios—Cataloochee Ranch Collection

Preface

I recently re-read, after many years, *Our Southern Highlanders,* that astonishing book by the Yankee librarian and historian, Horace Kephart. In 1904, Kephart simply abandoned his job and family up north to take up a primitive existence in North Carolina's Smoky Mountains. For nearly three decades afterward, his writings enlightened America about what he characterized as an eighteenth-century culture that still survived largely intact and unnoticed in that region.

Like Kephart, my father, Tom Alexander, Sr., was a town-raised boy, born in the Atlanta suburb of Decatur. And like Kephart, once transplanted to the Smokies, he adapted to mountain ways and attitudes so thoroughly as to become not only accepted and respected but even loved by the native society, which for all its overt courtesy and tolerance privately regarded outlanders—unable to hew a dovetail, track a bear, cure a ham, run a still—as comically incompetent. (The derision was mutual, of course, though city folk tend to be much more open in ridiculing mountain folks' ignorance and backwardness.) Even today, some twenty-three years after my father's death in 1972, I still encounter local people, strangers to me, who go out of their way to tell me what a fine man my father was. I suppose it all has to do mainly with the generosity, openness, and general integrity that he displayed in his dealings, together, perhaps, with a certain frank impetuosity that seems to resonate well in this region.

By the time my father entered the mountains in the early twenties, I suspect that the native culture had advanced a good way from Kephart's eighteenth century toward something more like the nineteenth. Certainly this was true by the early thirties and my own childhood in the Cataloochee Valley, a remote community of a few dozen families, largely housed in log cabins, devoid of electricity, telephones, or automobiles except, I believe, for the single car my parents possessed. Possibly my very first reliable memory is of dashing up the valley through the snow and across the frozen creek in a one-horse locust-runnered sled—at that time, the hill man's counterpart of a farm wagon—laden with rocking chairs, relatives, and neighbors, cowbells jingling.

Because we eventually moved to the even more remote location of the present Cataloochee Ranch, I spent only a brief time in mountain schools. The rest of my schooling was acquired in various boarding arrangements with far-off relatives or local private schools. Later on, the

Jane Alexander

The Cataloochee Divide looking toward Jonathan Valley viewed from Hemphill Bald.

first literary evocations of the mountain society I had known as a child—the laconic humor; the blend of self-reliance, independence, cooperativeness, and integrity; the casual regard for hardship and danger—were those I found in the frontier memoirs of Mark Twain and other nineteenth-century authors. Until I finally departed for the big cities in the fifties, I pretty much spent my youth swapping back and forth between the nineteenth and twentieth centuries, not altogether at home in either.

Nowadays, except for remnant peculiarities of language and certain traits and skills among the elderly, the average southern Appalachian community doesn't seem all that different from any other in twentieth-century America of similar class and circumstance. Radio, TV, automobiles, machines, and relative prosperity have seen to that. Kephart remarked that he had virtually never encountered a fat mountaineer, and he speculated that the explanation might be genetic; nowadays, overweight probably outranks croup, bloody flux, and milk-sick combined as threats to health in the region.

For all the modern ways, though, enough frontier-style self-sufficiency still resides in these mountains to inspire lots of disaffected outland youth seeking authenticity through traditional handicrafts and rural pursuits. The popularity of regional craft schools, log houses, and the Foxfire series of books on Appalachian skills and lore testifies to that.

Besides chronicling one man's love affair with an outdoor region, this book serves as a post-Kephart snapshot of a society in transition. My father composed it over many years, with some material based in articles he wrote for various newspapers and magazines. Some of the later material was written or dictated during his long incapacitation following a stroke in the mid-sixties. Much of that in particular has been edited by my wife Jane and me, not only to eliminate material that seemed redundant or extraneous, but also to interconnect some incidents and activities or put them in context.

Perhaps the main shortcoming of the book is that my father never got around to finishing it. In about 1960, for instance, he and my mother took it into their heads to start a ski resort—the first south of Virginia—on the ranch property. The main motive, as with many things they did, was to provide wintertime employment to descendants, ranch hands, and the local community. He became so absorbed in the intricacies of climate and weather, ski-slope locations, ski lifts, the new technology of snow making, and so forth as to find little time to write about it. Nevertheless, the Cataloochee Ski Area, which opened for business in 1961, pioneered what is by now a major wintertime industry in North Carolina and some adjacent states. By the time of Dad's death in 1972, the management of the ranch and ski slope had largely reverted to my sisters and their husbands, Judy and Rick Coker and Alice and Tom Aumen, with my mother, "Miss Judy," playing a slimmer role than formerly.

Several people who have read the manuscript have noted that Miss Judy figures remarkably slightly in it even though my father frequently acknowledged the obvious fact that Miss Judy really ran the Cataloochee Ranch, holding both the resort and farming operation together all the time he was off making a living as a professional forester. My parents were the closest and most interdependent married couple I have ever known, and I do not remember them ever exchanging a cross word, though surely they must have. My sister Alice has volunteered to remedy what we take to be Dad's oversight by writing a daughterly appreciation of Miss Judy. And sister Judy has compiled a profile of our father himself, relying largely on anecdotes from friends and former ranch guests.

Tom Alexander, Jr.
Cataloochee Ranch
September 1995

James Hannah, a descendant of one of the first white families to settle in the Cataloochee Valley. He served as a magistrate of the district. Other members of the family include Mark Hannah, park ranger in the valley for thirty-one years, and many talented musicians.

George Masa—Plateau Studios—Cataloochee Ranch Collection

Until fairly recently, self-sufficiency was a defining attribute of Appalachian mountaineers. Remote from specialists and complex supply systems, people were brought up to turn their hands to any task and to supply their needs from things at hand. Honey, fruits, and berries substituted for sugar, as did sorghum molasses rendered from cane. In these photographs, the mule or horse treading a circle at the end of a pole powers a mill that squeezes stalks of cane. The extracted juice is boiled down into the dark, heavy molasses with which people sopped their corn bread.

Margaret Morley—Courtesy of the N.C. Division of Archives and History

George Masa–Plateau Studios–Cataloochee Ranch Collection

As perfectly composed as a work by some old Dutch master, this washday scene by George Masa captures the sylvan serenity of the mountain life of seventy-five years ago.

Courtesy of the Ewart M. Ball Collection, Ramsey Library, UNCA

Cataloochee Creek, where the Alexanders ran their first guest ranch, has been renowned for many years as a prime Smoky Mountain trout-fishing stream. The valley through which it runs, once heavily settled, is now a favorite camping area for national park visitors.

Jane Alexander

The Plott Balsams viewed from Hemphill Bald.

Looking east toward Jonathan Valley in the Cataloochee Divide.

Jane Alexander

Introduction

For me, the craving for mountains is an incurable disease, comparable to the craving for whiskey by an alcoholic. At times I become temporarily surfeited, but when kept out of the mountains too long, I actually become mentally and occasionally physically ill.

I was not born in the mountains. I saw my first mountain when I was seven years old and traveling with my father from Atlanta to Clayton, Georgia, by train. He had promised to show me a mountain. To this day, I can recall the thrill I got when he pointed out to the right of the train and said, "That's a mountain."

It wasn't much of a mountain, actually, standing sort of ragged and blue-brown a couple of miles off. But that pitiful little hill, the last convulsion of a dying ridge tapering down into Piedmont Georgia from the Blue Ridge, may first have infected me with mountainitis, though it must have been a mild infection then. For I did not get on a mountain and look off. I did not meet a mountain man and hear him talk. I did not wade a mountain stream or watch a storm roll up and drench me and pass on across the ridges. I did not eat a ramp or taste corn whiskey. I did not kill a rattlesnake or chase a bear. I did not sit on a bed of deep green moss or sweat on a climb through a virgin forest. I did not see fall, nor feel spring slide her pastel quilt gently up the slopes and tuck them in. I did not saddle a horse in the woods at dawn. None of these just then.

Yet it seems now that my every move from that day on was pointed toward the mountains. I recall announcing shortly afterward that I was going to be a naturalist. The naturalist part was eventually modified to forester. It might just as well have been geologist, or engineer, or botanist. It wasn't a career that called me—it was a region.

If the sight of that first little hill did not give me the original infection, then at least it surely started the reactivation of a case lying dormant in my blood and inherited from generations of ancestors back in the hills of Scotland. My case is active now and definitely incurable. I am describing the development of the case for what it may be worth to younger people tempted to expose themselves to the malady. I must say, however, that not only is it not necessarily fatal but may even have its compensations. I have found it not too unpleasant.

George Masa—Courtesy of the Ewart M. Ball Collection, Ramsey Library, UNCA

A wintertime road in the mountains.

Snakes, Fires, Tourists, and Other Mountain Hazards

In 1913, when I was twelve, my older brother Miller, our friend Eugene Hicks, and I went on a camping trip up to a little mountain valley that a foster uncle owned. It was on Betty Creek, between Pickens Nose and Ridgepole Mountain in Rabun County, Georgia, just below the North Carolina state line, about eighty miles from home in Atlanta. There was a small artificial lake of six or seven acres with four or five cottages around its shores.

We arrived by train in Dillard, a tiny mountain village on the Tallulah Falls Railroad, early in the afternoon. We were directed to old man Dillard who ran the only hack. I'm not sure whether the village was named for him or for one of his ancestors. He wanted six dollars for the six-mile trip up to the lake, so we walked, carrying our blankets in shoulder rolls and our food and clothes in old suitcases. We reached our camp, an old lumber shack near the lake, just before dark.

We stayed a month. We fished with worms in the lake for perch and catfish and suckers and caught them. We fished with grasshoppers in the streams for trout and caught them. The men who owned the property did not seem to care much about fishing themselves, but once three of their friends came to visit and wanted to fish. The first day they fished for trout with fancy rigs but didn't catch many worth keeping. We were fishing for trout too, and caught some, including a fourteen-inch beauty I got on a grasshopper. So the visiting fishermen hired us to catch grasshoppers for them and paid us with artificial flies, but we still did better with grasshoppers.

The grownup mountain boys in the nearby hollows took a liking to us city kids, and on Sunday afternoons they would collect at our shack and play the pokerlike card game, set-back, for "a drink and smell." The winners of each game got drinks of raw corn liquor, while the losers got just a smell. It was the first experience we city kids had with mountain whiskey, and it was too strong and throat-searing for us to take more than a sip at a time, but by the time we quit in late evening all the mountain boys were a little drunk.

One Sunday the mountain boys came with a number of hounds and took us rattlesnake hunting up to the Steer Cliffs rattlesnake dens on the face of Ridgepole Mountain. The cliffs, which got their name from a steer that had fallen over them years before, were really steeply slanting bare rocks. We could climb them by catching our toes in the crevasses or stepping on bunches of grass growing out of rock crevices. The snake dens were under large, flat boulders lying about on the cliffs.

On the way we found a big yellow rattlesnake stretched out across the trail, unmoving. Right there the mountain men showed us how to kill a rattler with the least possible bother and danger. One of them stepped off the trail and cut a switch, a stout flexible sprout some five or six feet long. Approaching from the uphill side where the snake could not strike more than a foot or so, he brought the switch down sharply across the reptile's neck. This completely stunned the snake and it was easy to finish him off with additional blows. I have used this technique to kill hundreds of rattlers since then, for there is little chance of harm either to man or to the meat of the snake, if one is interested in rattlesnake meat. Many people try to kill rattlers by throwing rocks at them or by shooting them with a pistol. In most such cases, the snakes get away. We cut this big fellow open and he had a chipmunk—or ground squirrel, as the mountaineers say—inside him, barely dead. Miller got the snake skin for a souvenir.

Then we went on to the cliffs, the hound dogs sniffing at the big loose flat rocks lying about. From time to time one of the dogs would pause at a rock and bark and the other dogs would join him. We would turn the stones over and there would be one or two big rattlers under each. Dogs trained to hunt rattlesnakes!

We, of course, killed the snakes. We got, I believe, six grown ones, some brightly colored and some black. Two had young unborn snakes in them. The little snakes were in individual sacs like pale egg yolks and could move when we broke open the yolks; they were evidently about ready to hatch. I now wish we had noted whether the young were found in the yellow or black rattlers, for I still don't know whether color of a snake is really an indication of sex as many contend.

Our native friends took us on a two-day trip back into the big Nantahala Mountain Range to Standing Indian Mountain. The first day we climbed over the Nantahala Range, through Betty Creek Gap, and dropped down a mere trace of a trail through a magnificent virgin hardwood forest of oak, maple, chestnut, basswood, birch, and beech. We camped for the night on the bank of the beautiful Nantahala River, which, there near the headwaters, was only a small creek.

Chestnut! Green and luxuriant in those days, it made up more than a third of the forest. It was still untouched by the terrible chestnut blight that within twenty-five years was to wipe out this magnificent American tree. But when I was first there the chestnuts formed a rolling green canopy instead of the gaunt silver skeletons that later towered among the lesser greenery of those mountains.

>>> *Hardwoods typical of the middle levels of the southern Appalachians, as they existed in the early part of this century. Due to the combination of elevation changes, abundant rainfall, and the fact that the glacial ice-sheet pushed northern vegetation into these mountains but never covered them, the Smokies are said to contain a variety of plant species virtually unmatched in so small an area anywhere else in the world.*

Courtesy of the Ewart M. Ball Collection, Ramsey Library, UNCA

That first evening we caught fifty or sixty little black and gold brook—or speckled—trout, most of them less than six inches long. Along with cornmeal flapjacks, sorghum syrup, and coffee made in a large bucket, those constituted both supper and breakfast for us. The dogs had flapjacks too, plus the entrails, head, and feet of a groundhog. They had treed the hog in a leaning sourwood tree near camp, and one of the mountain boys climbed up and shook him out among the hounds to see the fight. The groundhog gave a good account of himself against the three dogs before they tore him to pieces. We saved the edible portions for our second night's supper.

We had a blanket apiece and lay in a damp, dark flat in the doghobble (leucothoe) on the river bank. Next morning we took a dim, brushy stock trail along a leading ridge from the river and topped out on the crest of Standing Indian about noon. The top was a long, narrow, grass-covered ridge, almost free of trees. Cattle, horses, and some sheep were grazing along the backbone and promptly approached us, either from curiosity or in hopes of salt.

Well before the white settlers came, livestock belonging to the Indians had grazed these Appalachian peaks in summer, away from the heat and insects of the lower valleys. This grazing, together with a little clearing and burning, first by the Indians and later by whites, to open the woods so they could shoot turkey and deer, probably accounts for the puzzling mountain balds so common in these parts.

The primary trees now growing on the higher stretches are oak until one ascends into the spruce belt of the Smokies and other heights, such as Grandfather Mountain, Mount Mitchell, and Roan Mountain.

Just as an injured fruit tree striving valiantly to pass on its life will often bear more fruit than healthy trees, those twisted, dwarfed, storm-torn oaks and chestnuts on the tops produced more acorns and nuts than their crowded brothers in the sheltered coves below. Thus, on the tops, there was and is mast, or tree-borne forage, for animals both wild and tame, more abundant than in the deep woods of the coves.

Wildlife gravitated to the parklike groves when acorns and chestnuts began to fall, and the hunters followed, both to kill game and to check on their livestock and salt it. Wild and tame animals intermingled, as they still do today on the western ranges.

We stood in the cool sunlight on the rocks where the Standing Indian fire tower is now—there was no tower then—and looked south over the sea of mountains in North Georgia: Tate City Gorge, Coleman River, the great basin where Burton Lake is now, on to Tray Mountain and Brasstown Bald in Georgia; west to Chunky Gal, Vineyard, Tusquitee Bald, Valley River, and the Cheoah ranges; north along the Nantahala backbone to Siler, Wine Spring, Wayah, the Burningtown Balds, and the distant Smokies; eastward back over Ridgepole toward the Fishhawks and the fascinating Highlands country. The view from Standing Indian is magnificent.

Some ten years after my first trip to Standing Indian, I was to spend many nights there. The U.S. Forest Service built a rough cabin on the spot, laboriously sledding in

Margaret Morley—Courtesy of the N.C. Division of Archives and History

One of the most rugged peaks in the southern Appalachians is Grandfather Mountain, located in the Blue Ridge near Linville and Boone. A favorite North Carolina scenic area since the nineteenth century, it has been imaginatively developed into a major tourist attraction since the 1950s by Hugh Morton, a descendant of the family that has long owned the mountain. Morton has also capitalized on the mountain's peculiar geology and large number of rare or endangered animal and plant species to convert the property into a nationally recognized center for research, conservation, and education.

from Black Gap the flooring and furnishings. As a forest ranger, I often stayed in the hut, with its log braces on all four sides butted down into rock crevices to keep it from being blown off the mountain. I learned to love those moaning and shrieking night winds that shook the cabin and froze the water in the cabin bucket despite two fires, one in the cook stove and the other in the conical Sibley heating stove.

On my initial visit we stayed at the top only two hours or so and en route back to our camp, the dogs treed another groundhog, way back under a big, flat rock. We could barely make out the shape of his head wedged in a crevice. The dogs couldn't or wouldn't bring him out. Our friends told my brother Miller to wrap his hand in his shirt and reach back under there and pull him out. Miller tried it, and the groundhog bit completely through his shirt, thumb and all. So we left that groundhog alone and went back across Betty Creek Gap to our home camp.

One afternoon a summer rainstorm brought the creek down high and muddy, dropping brown silt into the lake. When the rain let up, we went out fishing in a little flat-bottomed boat. We tied the boat to a stick stuck in the mud where the stream emptied and made a huge haul of suckers, catfish, and one or two trout, using worms for bait.

Lex Darnell, the camp caretaker, took us bullfrog hunting on the lake one night. We shot frogs with a .22-caliber rifle, shining their white throats with a flaming lightwood pine knot carried in a calf muzzle nailed to the top of a stick. We got plenty of them—twenty or more—and ate backs as well as legs. Also, we learned how hard it is to kill a bullfrog and keep him dead.

The flat pastureland surrounding the little lake was dotted with low alder bushes. A young bird, probably a robin, which had evidently just left its nest, was trying its wings, flying from bush to bush. Slithering behind him at astonishing speed was a blacksnake, carrying his head about twelve inches in the air and trying to catch the bird. The young bird, evidently knowing the danger from instinct, would alight in a bush and the snake would almost reach him. Then the awkward bird would fly on another twenty or twenty-five feet to another bush; and the snake, head high to enable him to see the bird's flight, would be right behind. I did not see the final outcome.

Blacksnakes, of course, are harmless to humans, but Lex Darnell described an encounter he had with one. He was walking down a road with wagon ruts and came upon a blacksnake with its head down in a rut, apparently drinking water. Lex walked up and kicked the reptile into the rut. The snake reared up, weaving its head in the air and looking right at Lex. Lex stepped off a few feet and started to reach down for a stick or rock, but the snake started toward him. If a snake's eyes can be said to blaze, that snake's eyes blazed. Every time Lex stopped, the snake would move toward him. Lex finally got a stick, whereupon the snake crawled off into the bushes, figuring it was no match for a man with a stick.

I have come upon a blacksnake lying in dry leaves and twitching the end of its tail in the leaves to make a rattling noise not unlike a rattlesnake. How does a blacksnake know that human beings are afraid of rattlesnakes?

At the Betty Creek camp we had our own spring box in a little laurel-bowered seep close by our cabin shack. Our mountain neighbors gladly sold us all the milk, buttermilk, and butter we needed. At various times, however, friends made their way up to visit, so our larder had its ups and downs. Late one afternoon, we learned by mail that we were to have a guest. I borrowed a horse from a neighbor and rode the six miles alone to Dillard for supplies. I came back after dark, and I still have a clear memory of the bravery it took.

After a month of this camp life, I was sitting on a rail fence with the other boys one day when a hornet hit and knocked me off. I cried. The boys chided me for it, and the next day I went home to Atlanta alone. No doubt homesickness had as much to do with it as the hornet or the chiding, but I had had a month in the mountains and the mountain fever was in my blood.

✣ ✣ ✣

I left school in the spring of 1918, in the end of my third year in high school, to join the Marines. After boot training at Parris Island, South Carolina, in the almost intolerable heat of the swampy coastal region of that state, and a tour in the Savage Arms plant in Utica, New York, in late summer, I was made machine gun instructor at the Marine base in Quantico, Virginia—a kid of eighteen training veteran Marines in the intricacies of this new weapon. We lay on ponchos on the sandy banks of the Potomac, tearing down and putting together four makes of machine guns and automatic rifles. The tough old Marines were excellent students, so well disciplined that they gave no trouble at all to their very young instructor.

In late October of 1918, we were formed into a special machine gun battalion and shipped to France, landing at Brest on November 9, two days before the armistice. I enjoyed almost a year in France, returning to the States in August of 1919 in time to enter high school for my senior year. The following fall, I entered forest school at the University of Georgia in Athens, Georgia.

Our student body consisted largely of veterans attending school under a rehabilitation program in which tuition and board were paid by the government. Most of those men had little interest in learning, being chiefly interested in "riding the government gravy train," as they called it.

A requirement of all forest school curricula is that students must attend summer training to put into practice the things learned in books. Our school had no summer camp of its own, but a professor offered the use of his farm in the North Georgia mountains. In the spring of 1921, a "rehab" student named Brown and I were sent up for a weekend to look over the site and decide what would be needed to make it usable for our summer camp. We set out for Blue Ridge, Georgia, the town nearest the campsite, in Brown's Model T. There were no paved roads, but we got along all right over the red and purplish clay roads until a few miles above Dawsonville, where grades got steeper and steeper and washed-out ditches deeper and deeper. Late in the afternoon the old car began to falter, and just before dark, the rear end gave one last clash and tore up. No houses were in sight, but we remembered passing one about two miles back. We walked back and found it occupied by a hospitable mountain man. He promptly invited us in, and we spent two nights with him and his cheerful family while our car troubles were taken care of at a little roadside garage a couple of miles down the road where our car was pulled by mule.

Our hospitable host was a preacher on Sundays and a farmer the rest of the week. He lived in the heart of moonshine country and told us that if he fired a gun from his front porch, the sound could be heard at seven stills tucked away in the wooded hollows. He was broad-minded about it all, never interfering with the stills or reporting them to the authorities. Many of the operators were friends and some came regularly to his Sunday church services.

We did not reach the camp that weekend, as we had to return to school for Monday

Margaret Morley—Courtesy of the N.C. Division of Archives and History

An early twentieth-century whiskey still. The main change in the technology of illegal whiskey making between then and now is that nowadays most of the components are likely to be made of stainless steel rather than wood and copper.

classes. But when summer came, our school made camp at Cooper's Creek in Fannin County in the upper reaches of the Occoa River basin of the Blue Ridge range.

The basin was inaccessible except by treacherous little dirt lanes through the red clay, always impassible in winter and in the summer when it rained. Yet, the basin was fairly thickly settled and is the point of origin of some of the finest apple brandy to be found anywhere in the southern Appalachians. Some of the same families who were making it then still do so, illicitly, of course. Some make good corn whiskey too, but they do not pride themselves on the whiskey as they do on the brandy. Too often nowadays, the corn whiskey is made with too much sugar.

The rehabs at the camp were a tough crew. Our camp director, Professor Burleigh, had a rule of no drinking and no poker playing in camp. Once several rehabs begged off from an overnight hike to Three Forks Ranger Station some twelve miles away on grounds of sickness. When we got back to camp next day, earlier than expected, we found them playing poker. Professor Burleigh sentenced them to KP. Considerable grumbling ensued, and one man named Rowan got his knife out and began sharpening it, saying he was going to cut his initials on Burleigh's heart. That didn't worry Burleigh much, but young Holland, a delicate, sensitive student, voluntarily took Rowan's KP turn, quieting the trouble.

We had a Negro cook, and supper wasn't just right one night. Shortly after some noisy threats about "lynching a nigger," the cook disappeared. He hiked the twenty miles to Blue Ridge, Georgia, that night and caught the train back to Athens.

After the cook's departure, we did most of the cooking, baking barely edible wheat bread in a three-legged Dutch oven. Once the boys shot a half-grown sheep and we had fresh mutton. The owner suspected us, but we had buried the head, hide, and entrails. He couldn't prove anything, but we bragged too much for complete safety.

Two of the bunch got drunk one afternoon and began arguing. They were going to shoot it out, or rather, one was going to shoot and the other cut. But when we urged them on, they subsided.

Part of our course consisted of helping Dale Edmundson, the government ranger, build a trail to the top of Licklog Mountain. I was made straw boss of a gang of students, camping in a couple of tents on top of the mountain. We didn't have cots but slept on pole beds covered with dead leaves and plenty of chiggers. We didn't build much trail.

One Sunday we all went down the other side of the mountain to a valley church, to an "all day singing and dinner on the ground," and I never saw so much fried chicken and pie in my life. Some pretty girls were there too, mostly with boy friends, but one or two of my gang found unattended ones and went down again later, courting. I guess I felt my importance as foreman and didn't make any girl friends, or maybe I was just bashful.

I shot a rattlesnake coiled under a flat rock near the falls we used for a shower bath. Later, when I took the boys to show them the rock, we turned it over and found a copperhead coiled under it. We put a forked stick over him and caught him alive, but he twisted himself into such knots that he tore holes in his own skin. I have often wondered since if the mountaineers are right in claiming that a copperhead is a rattlesnake pilot and stays around rattlers.

Lee Anderson, a rehab who had been raised in the North Georgia mountains displayed the greatest skill I have ever seen at catching little speckled trout from tiny, laurel-bowered mountain branches. Lee was a big, quiet, smart, dark-skinned fellow getting a free education because he was blind in one eye, the result of an accident with a quirt when he was in the cavalry in Texas on the Mexican border in 1916.

To fish, he used about three feet of leader on a short pole and stick bait, a kind of insect larva found under water in tiny cocoons made of twigs. Lee would slip silently and invisibly down over the moss-covered rocks through the laurel along the stream. He would drop the bait in a ripple at the head of a dark pool, and quick as it touched the water, a speckled beauty would snap it, and Lee would instantaneously snag him. Timing, it was, and experience. I was too slow, I guess. It was years later before I learned to do it nearly as well as he.

The storekeeper/postmaster at the community of Margaret had a barrel of corn whiskey buried behind his corncrib. A little rubber hose reached down into the barrel and the postmaster charged a dime a suck. Good suckers, able to handle the raw liquor straight without gagging, got their money's worth, but greenhorns like me didn't fare so well. Even so it was there that I first began to like mountain corn whiskey.

After summer camp was over, Professor Burleigh and I hiked the hundred miles to Highlands, North Carolina, taking five days. Burleigh was looking for specimens of birds' eggs. We spent the first night at Young Harris, Georgia. Next day at noon, having hiked twelve miles, Burleigh bought a new pair of shoes at Hiawassee, Georgia, and that afternoon and evening we walked twenty more miles. Burleigh's feet were covered with blisters but he didn't complain. I was nearly dead by nightfall, however, and every hundred yards or so, I had to lie down in the middle of the road to rest.

At about ten o'clock that night we got to Tallulah River where a high bridge over Burton Lake was being constructed. A ferry was being used until the bridge was finished. Seeing a light across the river, we yelled and succeeded in attracting the attention of the ferryman, who came and took us across, but wouldn't let us bunk in his shack with him. He directed us a quarter of a mile up the road to a construction camp where the Negroes that worked on the bridge were staying. It was Saturday night and the workmen who hadn't gone into the nearby town of Clayton were drunk, but the mulatto cook was half sober. When we discovered conditions in the camp, I slipped out of sight and strung a leather strap over my shoulder under my shirt to make it look like the strap of a gun holster. We told them we were government men examining the timber, and the cook fixed us up with a good chicken supper and bunked us on two of the vacant cots. I slept that night.

The next day we walked on into Clayton and down to Wiley, Georgia, to visit my brother Miller who was a counselor at Camp Dixie for boys. We then went back to spend the night at a hotel in Clayton. There, my father joined us. The next morning, the three of us started out on foot for Highlands. After we had walked four miles up the valley road toward Dillard, we discovered we had left our hatchet at Clayton, so Burleigh walked back the eight miles round trip to get it, while my father and I continued on to Dillard. That Burleigh was a walking fool.

We planned to catch the mail hack at Dillard that afternoon for Highlands, but the fare was too steep for Burleigh and me, so we sent our packs on with Dad in the mail hack while we walked. We took the shortcut trail by Mud Creek Falls, the same falls that William Bartram wrote about in *Bartram's Travels* in about 1770. By taking shortcuts, we got to Mrs. Laird's mountaintop camp between Scaly Mountain and Highlands almost as soon as the T-model mail car did over the muddy road. We stayed a few days around

Courtesy of the R. Henry Scadin Collection, Ramsey Library, UNCA

Looking west from Mount Toxaway, near the North Carolina-South Carolina-Georgia corner, toward Cashiers and Highlands. This region has long been a summertime watering hole for visitors from hotter lands to the south.

Highlands and climbed Scaly, Rabun Bald, Satulah Mountain, and saw Glen Falls and other interesting places in the Highlands vicinity.

Highlands in those days, as now, was popular as a summer resort even though there were no paved highways or golf courses. Yet it was beautiful, with long, rambling white hotels in the dark pines and hemlocks. The woods and the glistening white sand roads always seemed moist and cool.

School money got scarce in our family, so in March of my second year, I left college and took a job as a rodman on a survey crew surveying lands in Virginia for acquisition by the U.S. Forest Service. Neville (Tod) Sloan was chief of our party. Our crew gathered at Bristol, Tennessee, and proceeded by railroad through Pulaski to Ivanhoe, Virginia. There we set up camp about seven miles back in the mountains. It was March, and cold, and one night a storm blew the big cook tent down on our cook Posey Fouts who was trapped in his cot until we rescued him at daylight.

One day we ran a survey line directly over the top of a liquor still. The operators were gone but the still was hot, with liquor dripping from a copper worm into a gallon jug, another full jug nearby. One of our axemen took the full one, but the liquor was too hot to drink.

We got into an argument as to whether a bull bat and a whippoorwill were the same bird. One evening I decided to prove my contention that they were not. I crouched by the rail fence near our tents where a whippoorwill often perched, caught him alive, and showed him to the boys to prove he was different from a bull bat. My, I thought I was smart.

The region we were surveying had been logged years before for sawtimber. Even the young trees had been stripped to make charcoal for the iron ore smelters at Ivanhoe. It was extremely unattractive country. As we finished surveying in May, though, we came out into the lovely bluegrass limestone area around Cripple Creek, Virginia, where the cattle were sleek and fat and the saddle horses gaited and plentiful. One wanted to lie in the cool green pastures or maybe even buy a farm and settle down. My first sight of mountain cattle country had started me dreaming about mountain grass pastures for my own cattle.

Our next surveying job was to be in the Nantahala Forest area, and our headquarters was at Franklin, North Carolina. I was stopping at the old Junaluska Inn Hotel at Franklin. My first day in town I met Mr. and Mrs. E. C. Kingsberry who lived in a beautiful house in the outskirts of town known as Pleasant Hill. I came down with the flu and the Kingsberrys insisted on my leaving the hotel and staying in their home until I was well. Soon afterward the crew boss, Tod Sloan, also became ill, so I had to take over the crew and complete the project. I had earlier been promoted from rodman to transitman and now was acting chief of party. Getting up in the world and proud!

One day our axemen out ahead of the surveyors encountered a rattlesnake den on a rock cliff, which we told some natives about that night. Next day they went up the mountain with us, captured eight snakes alive, and brought them down in a sack. While we were looking at them in the sack, one huge rusty fellow began shedding his skin. In his brilliant yellow and black new skin, he was a beautiful creature. The mountain boys took the snakes home, intending to try to sell them, I think.

That fall when the field work was over and we were working up our field notes at Franklin headquarters, the great Standing Indian fire broke out. It eventually burned fifteen thousand acres in Georgia and North Carolina, near where I had hiked as a boy. We all hiked into the woods that night and fought the fire for three days and nights.

Forest fires in those hardwood forests generally are only leaf fires burning on the ground, and the common fire-fighting technique is backfiring. Equipped with especially made rakes having teeth made of triangular-shaped mowing machine blades, we went out ahead of the oncoming fire and raked a path, then lit the unraked leaves on the fire side of the raked line. The two fires burned together, creating a burnt-over strip wide enough to prevent the main fire from jumping over. We built a fire line fifteen miles long between the Coleman and Tallulah rivers.

We laid out on the frozen ground one night to rest in a little lean-to of brush. I think I might have frozen if John Wasilik, a Forest Service employee, had not succeeded in keeping a fire going using green laurel and rhododendron. Nobody really slept, but we got some rest for our tired muscles.

During those three days and two nights I saw feats of incredible endurance, some approaching heroism. One crew who had been working on the fire line since Thursday had just gotten back up to the lookout cabin on Standing Indian at three o'clock Saturday morning when we saw the fire break over their line far down in the Tallulah Gorge.

Without hesitation, that crew took off down over the cliffs in the darkness to attack it before it could get out of hand. The party was led by Gilmer Setzer, the only one who knew the way down through the cliffs in the dark.

An hour or so after Gilmer's party left to fight the break-over, two thousand feet down in the timbered gorge we could see their little back-fire blaze spring up and slowly creep around the oncoming fire. By daylight they had it cut off.

Gilmer was tough like many mountain men I have known since, but even he took a few days to recuperate after that fire. We later learned that he had stuck a pitchfork tine through the calf of his leg the first day of the fire, and by Saturday his leg was blue and badly swollen. But nobody knew anything about it until the battle was over. His coat had also burned up in a bush where he had hung it and he fought on without one, with the thermometer around ten at night and bitter winds blowing.

✣ ✣ ✣

Federal forest acquisition funds ran short that winter, leaving me out of a job in the mountains, and I went back to Atlanta for a couple of months as Secretary of the Georgia Forestry Association. But in January, Ranger Brown resigned his job at Franklin and it was offered to me on a temporary basis, pending my passing the civil service ranger examination, which was to be held in March. After less than two months in the flat country, my mountain fever had flared up. I accepted by fastest return mail, without stopping to inquire about salary or any other conditions. Reaching Franklin again, I bought a horse from Tom Angel, the livery stable operator. My purchase was a handsome sorrel stepper, an ex-cavalry horse. I named him Sparkplug.

My first duty was to get acquainted with the extremities of the ranger district in the Wesser and Tellico portion of Swain and north Macon counties, near Nantahala Gorge. I rode through in the dead of winter by way of Burningtown, Tellico, Needmore, the Tennessee Gorge, and the Nantahala Gorge, stopping the first night with Robert Ramsey on Tellico Creek. He had a big two-storied white house, which he had built entirely by himself. He ran a little country store, had an orchard, bees, cattle, and a rocky but productive farm tucked back in the mountains, almost impossible to reach by car except in the summer. He and his family were superbly hospitable as were all mountain people. The second day, I rode on to Wesser via the Little Tennessee Gorge and spent the next night with Bob McMahan at his house on the banks of the Nantahala River.

My next stop was Coon Cochran's at the head of Queens Creek. Coon was six feet, six inches tall, weighed about 250 pounds, had a voice resembling a lion's roar, a black beard, hands like hams, and tales as tall and vivid as any ever concocted about Paul Bunyan. Nothing he owned and nothing he talked about was commonplace. My stay at Coon's included my first night in a true featherbed that flowed up around me like soft dough. But my most vivid recollection about this man living so far back in the mountains that he rarely got to town more than once a year was that he had successfully grafted limbs from one apple tree onto those of another. He had two kinds of apples ripening on the same trees. Coon also taught me the delights of mixing sorghum syrup or wild honey with butter, mashing it all together with a fork and spreading it on hot corn bread. Afterward, you wiped the plate clean, for politeness, with the last piece of bread.

On inspection trips to lookout cabins, I often tied my fishing rod on the back of the saddle and stopped to fish awhile in the beautiful, tumbling little streams. Sparkplug got used to being tied up in the woods for long periods alone.

On one trip to Standing Indian, I had enough horse food with me for one night. I tied Sparkplug outside the lookout cabin on the leeward side of the cabin. Gilmer Setzer, the lookout man, and I had supper and spent a couple of hours jawing with the other fellows by phone on the one-wire "tree" line before going to bed. During the night a bitter blizzard blew up, and when I went out next morning Sparkplug was white with the frozen fog and shaking all over. I spread a tent fly over him and tied it on as best I could.

After breakfast Sparkplug and I started off the mountain through driving snow, since I had no more feed for the horse and no place to shelter him. By the time I reached Deep Gap two miles below, I was numb with cold, so I got off and led him. The storm kept on and when I got to the little dirt-floored stock-hunters' cabin at White Oak Bottoms six miles from the top, I had to stop and kindle a little fire to thaw out.

I decided then to turn off the mountains at Wallace Gap and head for the settlements, but for awhile the sun came out deceptively, and I decided to carry out my original plan and go on through the mountains to Wilson Lick Ranger Station on Wine Spring Bald. Two or three miles beyond Wallace Gap, the storm closed in again and it became evident that I was in danger of freezing to death.

Sparkplug did not lead well ordinarily and with the blizzard blinding him, I just couldn't make him follow. I could not travel fast enough to keep the blood circulating. Knowing that he knew that trail as well as I did, I finally tied the reins back on the saddle and, turning him loose, I set out alone, running to stir my blood. When I first left him, he looked surprised, but soon started on down the graded trail after me, slipping at every step. I took the shortcuts from switchback to switchback, and in a few moments Sparkplug was lost from sight.

I made it to Wilson Lick in the blizzard, arriving about four o'clock in the afternoon. We always had wood on hand at the station, and I quickly kindled a fire and thawed out. In about an hour, along came Sparkplug, ears erect, snorting steam. We had a little stable, with hay and oats in the loft, so both of us were all right. The blizzard subsided into clear, still, zero weather during the night, and next day we continued on through to Tellico Gap and Robert Ramsey's.

Many times during the winters, we would ford the Nantahala River or Burningtown or Tellico or some other flooded creek and the water would freeze hard on my boots and Sparkplug's legs and belly, but he never seemed to suffer any ill effects.

Sometimes in summer, to vary the monotony of long rides through the forest, I would hang the reins on the saddle and walk on ahead. Sparkplug would follow until he would reach a juicy patch of grass on the upper bank of the trail and would stop to graze until

Ranger Tom Alexander atop Sparkplug in the Nantahala National Forest in the early twenties. Traveling through the mountains in midwinter, Tom would often dismount and jog to keep warm, leaving Sparkplug to trail behind—when Sparkplug felt like it.

I was out of sight. Then, head high and tilted to one side to listen, he'd come stepping rapidly along to catch up. I started playing tricks on him. I'd hide behind a tree or in some bushes and keep still until he was past me. Ears forward, stirrups slapping his sides, he'd go by me at a brisk walk. Then I would whistle or laugh, and he would stop, acting nonchalant and trying to appear interested in the dead leaves or huckleberry bushes on the upper side of the trail.

Finally, he paid me back. I came down off the ridge into Tellico Gap ahead of him one day and turned down the old road to the right toward Robert Ramsey's and Franklin. Some two or three hundred yards down the road, I hid and waited. Five minutes, ten minutes passed and no Sparkplug. Then I went back up to the gap and there I found his tracks. Instead of following me down to the right, he had taken the left road to Coon Cochran's, three-quarters of a mile below the gap.

On another afternoon, this time in summer, I had completed my round of the lookout towers on the high peaks of the Nantahalas and turned my horse's head down toward the headquarters in the valley at Franklin. Our downward trail through the forest widened finally into a steep, rocky road and the first shacks and cabins came into sight. The first of these was particularly dilapidated and weatherbeaten and was surrounded by the remains of a hand-split paling fence.

In the doorway stood a lean, black-haired mountain woman. Outside the fence and across the road a boy about six years old was playing. The woman spied me astraddle my high-stepping sorrel, headed for home and champing at the bit.

Fearful for the child's safety, the woman yelled out, "James! James! You come in this house!"

"I'll not do it!" the six year old shot back.

"I'll make you think not do it, young man! I'll whup you till you jump up and down and holler, 'Hell! Hell!'"

But I couldn't laugh until I got around the bend in the road.

✣ ✣ ✣

In those days, many of the hardy mountain folk "raised" half-wild razorback hogs for their meat, their livelihood, their sport, and as an excuse to get away from home. The razorbacks ranged the mountains, including, by permit, the national forests under my care, and one of my duties was to issue permits, collect fees, and check on ranges to avoid overgrazing. There were hundreds of these permittees.

The razorback never saw the home of its owner until the fall he was led out of the mountains with a rope tied to one leg, fattened for a couple of weeks on corn, and slaughtered. The rest of his life, he sought his own living from the forest. This left his mountain owner pretty free of responsibility. Every spring, the hog farmer rode into the mountains with a poke of cold corn bread, fat-back bacon, and Pilot Knob coffee to resurrect acquaintance with his stock. All day long he tramped the brush, calling hogs, and reading the myriad signs of the woods—the game traces, man tracks, weather symptoms, and timber stands. But he also tended his neighbor's business, employing a leased wire on the grapevine telegraph, together with observation, deduction, memory, and a knowledge of the code of the ear-crops distinguishing each man's stock.

For example, George Roper, a stock ranger from Sugar Fork, might observe a sow rooting along the face of Chestnut Ridge. He would note the gaunt sides. "Sarn take," he would muse, "that old sow's got pigs!" Easing nearer he would read, "Crop and split in the right and a crop and underbit in the left. Half the new crop's been new cut."

Translated, this means that while the sow was still young, a portion of the right ear was sliced off and a slit made in the remainder; from her left ear, a right-angled piece was taken from the upper outside end and a moon-shaped section cut from the lower lobe. Later, however, another right angle had been hacked from the end of the left ear, leaving it cut straight across. After analysis is applied, it is obvious, of course, that the sow is the property of Dan Sweatman, bought of George Norton. George's mark, as is well known, is a crop and split in the right ear and an over-half crop and underbit in the left. By the simple process of taking off another corner of the hog's left ear, however, it was deeded over to Dan Sweatman, whose mark is a crop and split in the right and a crop and underbit in the left.

The message of the new family might then travel by way of Josh Slagle, George Moffit, Ves Dills, and Lige Norton to Dan Sweatman. Dan thereupon would see it his duty to make hasty preparations for a trip mountainin'.

"Mamie," he would say to his wife, "sack me up a poke of rations, and tell Ed to saddle the red-legged mule. That old sow I traded from George Norton has brought pigs out near the Chestnut Ridge camp, and I'm going to mark 'em afore some hog stealer beats me to 'em."

The majority of hog rangers were honest, but there were some who would place their marks on pigs originating in another man's drove. It was easy to steal hogs merely by changing their ear marks and easier still to mark newborn pigs. Still, the entire population of hog owners constituted a vigilante committee for mutual protection and assistance. Unlike the rustler and brand-changer of the cattle states, the hog stealer was seldom the victim of summary justice. He was usually tried in court, but it took an educated judge to understand the parlance of the hog rangers and to comprehend the chain of circumstantial and deduced evidence employed by the plaintiff.

Then, every fall, the overworked hog rangers had to go out again and loosen bell-collars. For fall was the time when hickory nuts littered the forest floor and sweet acorns from white oaks rolled under the leaves. Then it was that the normally slab sides of the razorback hog finally began to fill. Throughout the fall, the hogs swelled and swelled. By December, a half mile was a good day's wobble.

About one in ten of the hogs was a bell-hog, carrying a bell attached to a strap of groundhog-skin leather. And as the fall advanced, the collar would sink deeper and deeper into the folds of fat and finally disappear altogether, leaving only a bristle-lined crack to indicate a collar's presence. The bells' tintinnabulations became useless clicks.

At this stage, the mountain man made his extended trip into the wilderness to loosen bell collars and incidentally to hunt squirrels, pheasants, wild turkey, and coons. Coons were preferred. By day, this man hunted hogs and loosened bell-collars; by night he hunted coons.

By late December, the nuts were gone, and the man went out yet again and drove, led, or tolled in one or more candidates for killing. The killing was accomplished by the blow of an axe to the head. Then, the carcass was immersed in a barrel of water brought to the boil by means of large rocks heated in a fire. After that, the entire family took part in the scraping festivities, using hoes, scissors, and butcher and barlow knives as implements.

In my travels through the mountains on inspection trips, I would sometimes stop in one of the little stock-hunters' cabins for the night. Occasionally, a group of the stock owners would make the cabin on the same night, and we would share the shelter, bunking down together on the dirt floor. And there, lying on the ground in the black, unventilated, smoke-filled, windowless cabin, after a supper of onions, corn bread, fatback, and black coffee, with no light but the coals of the fire, I would hear stories

Courtesy of the N.C. Division of Archives and History

Since the earliest days, hunting has always been an important source of food for inhabitants of the steep and rocky hillsides of the southern Appalachians.

of such pure-spun imagination, keen repartee, and rich humor that when I awoke at the four o'clock rising hour, my sides still ached from laughing the night before.

✣ ✣ ✣

Most mountain tales originate, I imagine, in real events—bear hunts, bootlegging arrests, family squabbles, domestic animal episodes, and so forth—though tellers often find the need to add improvements. And certain individuals seem to become heroes of many stories, so that the mere mention of such a person will call forth an accumulation of anecdotes. One such individual was Sam Woody of Bryson City, whom I had known as a cook in several logging camps back in the big timber and who later came to work for me as a horse wrangler and woods cook.

Like many cooks, Sam could be temperamental, and it wasn't a good idea to be critical of his work. In one logging camp where he worked, the crew had gotten into the habit of saying uncomplimentary things about the fare. One evening Sam announced, "The next man I hear complainin' about the food in this camp can start cookin' hisself, fer I'm a-leavin'."

Then Sam started the meal. As he put the stew on and started seasoning, the top of the salt box came off and the whole boxful went into the stew. Nevertheless, Sam served it up with the corn bread. One hungry logger filled his plate and took a big mouthful. "Godawmighty!" he sputtered, "that stew shore is salty!" He paused as everybody else looked up, interested. ". . . And that's jist the way I like it!"

Todd Henry, a former logger who later came to work for me, once worked at the big camp on Bradley Fork, where Sam was the camp cook. Todd said that Sam had a pet rooster that followed him everywhere, even ate and slept with him. Every once in a while, Sam would have too much to drink and fall into his bunk before the rooster was ready for bed. During the evening Sam would wake up and spot his friend perched on the footrail of this bed. Depending on which way the rooster faced, Sam might yell, "Somebody please turn the rooster!"

According to Todd, Sam complained one morning that he hadn't had a mess of bear meat in a long while and that it looked to him like some of those trifling loggers could bring in a bear once in a while. Later on, Sam sent his helper Charlie out for firewood, and while Charlie was up in the brush chopping up a snag, he happened to step on a pair of young bear cubs trying to hide under a log. The cubs squalled, of course, and their old sow-bear mother came a-running. Charlie took out for the cook shack yelling, "Open the door, Sam!"

Sam looked out the door just in time for Charlie to run in, with the mad she-bear not far behind. Charlie headed for the back door hollering, "When she gets in, shet the door! I'm goin' back fer another'n!"

✣ ✣ ✣

Courtesy of the R. Henry Scadin Collection, Ramsey Library, UNCA

A prosperous valley farm, home of a solid citizen of the southern Appalachians.

The valleys and coves of the southern Appalachians are dotted with comfortable two-story painted homes surrounded by fields of fine bottomland yielding seventy-five to one hundred bushels of corn per acre. Herds of beef or dairy cattle spread over the uplands, while woodlots supply the wood needs of the farms. The owners are intelligent, hospitable, hardworking, God-fearing people; shrewd, independent, and humorous; citizens of a high order. Many of their children go to college, and a surprising number work their way into positions of importance in the outside world. The all-too-often drawn picture of the typical southern mountaineer as a lean, silent, shotgun-carrying moonshiner is largely false.

During Prohibition, and particularly before the Depression, a little whiskey was made almost everywhere in the mountains despite great activity on the part of both local and federal officers. But quantity production was concentrated in a few main areas, which supplied much of the demand of the large cities outside the mountains. Among these major areas were the Tate City-Coleman River section of Rabun County, Georgia; the Cosby area in Cocke County, Tennessee; the Dark Corner of Greenville County, South Carolina; and Wilkes County, North Carolina. The first largely supplied Atlanta and

Piedmont Georgia, the second supplied East Tennessee and much of Western North Carolina, and the third and fourth the cities of Piedmont North and South Carolina.

Moonshining still goes on but is centered mostly in a few areas, followed mainly by natives from a lower strata of the population. In a given valley most of the residents know who is making and often know about where the still is located, but unless the still or the operator become a nuisance for some reason, it is not reported. People, in the main, mind their own business.

A crusading individual may occasionally take it upon himself to break up the local moonshine industry. He might openly announce his intentions, but more often he secretly reports the location of stills to the law. Retaliation may take the form of a barn burned, a milk cow killed, or, rarely, bodily injury or even death to the tattler or a family member.

My ranger work included construction and maintenance of trails, telephone lines, and roads; stock ranging; leasing small tracts of farmland and houses the government had acquired; fire fighting and prevention; and timber selling. On the Nantahala, where timber selling was the major work, government policy favored the little man. We made literally hundreds of sales of tanbark, cross ties, chestnut telephone poles, "board trees," and sawtimber for less than $100. These small sales served to clean up the cutover areas of salvage material, preparing them for future full crops of timber, while providing the small entrepreneurs the means to piece out their meager incomes. At one time I had 120 separate sales to administer. It was my job to look over the area involved and estimate the amount of material; make out the contracts or permits, arrange the payments, and collect the money; scale, measure, or count the material cut; and check the areas frequently to see that cutting was properly done.

It was surprising how well the buyers obeyed the cutting rules written into the sale permits. With the rarest of exceptions, they lived up to their contracts so that we seldom had to invoke penalties or cancel sales for infractions. Throughout my experience in the southern Appalachians, I have found the overwhelming majority of the native people honorable and dependable.

Saturdays were town days for all the permittees and buyers, when all the little entrepreneurs came to my office to make their reports and payments or to apply for more timber. On these days my office would be packed from early morning until late evening. Regulations required that all payments to the government be made by postal money order or cashier's check. There would be long lines of men at the money order window of the post office, drawing orders ranging from $1 to $100 made payable to "Treasurer of the United States." I always made it a point to be in my office on Saturdays to attend to these business matters, usually spending the other five working days in the field on inspection.

After work on Saturday when the weather was good, I would often head for the mountains with friends to fish, either to the Black Place on the Nantahala or up on Cullasaja between Franklin and Highlands, in the gorge below Dry Falls. The old road up Cartoogechaye Valley across Wallace Gap to the Black Place on Nantahala was steep as a horse's face in spots and crooked as a dog's hind legs everywhere, but we would go across there on weekends in Bert Lyle's old Dodge. The car would overheat and quit, and we'd scotch her with rocks and wait for her to cool. Bob Davis usually brought along a glass and spoon for fishing tackle, and if we did not have the rest of the ingredients we could usually get them at the Black Place from somebody. Bill Moore was both the best and the most enthusiastic fisherman I have known, and I've often seen him bring in fifteen or twenty nice rainbows in an afternoon's fishing. He couldn't get enough fishing in daylight so sometimes we fished with worms at night at the big holes.

Dark Cowdung and Rube Wood were our favorite flies, but sometimes the rainbow trout would go for a Gray Hackle with a yellow body, Olive Quill, or Professor. For reasons no one ever explained, these patterns are never as effective on other streams outside the Nantahalas.

Cullasaja, Indian for "Sugar Fork," was even better fishing than Nantahala before the highway was built along it, causing mud and silt to wash in, but the old road up was a nightmare. Sometimes we'd have to leave the car and walk to the camp above Buck Creek. Beginning in the gorge just above the lower falls, on up by Henry Robertson's camp, through the rough by the "49" hole, and on up to Dry Falls was real fishing.

It was there that my most memorable trout fishing experiences occurred. One day five of us were fishing Cullasaja; the trout were sluggish and we had had little luck all day. It was almost dark, and Bill Moore and I were the only two left in the stream, I on a wide, smooth stretch of the river and Bill a little below me. The others had taken down their rods and were idling along the bank watching us. Suddenly the water exploded. Trout by the dozens—ten- and twelve-inch ones—began popping the water all around me. I heard Bill yell below, and a glance down the glistening river showed me that the same thing was happening his way and on beyond. The entire stream was alive, the trout leaping like bubbles in hard boiling water.

I was using two wet flies of different patterns on my leader. I landed doubles three times in succession and singles at every other cast. Trout would actually leap up and grab my flies before they could hit the water. The whole thing lasted no more than ten minutes, and in that time I caught fourteen nice fish. Bill had about the same luck. In the meantime, the astonished fellows on the bank jumped into action. Rods came out of cases, lines, reels, and fly hooks flew in every direction, as those poor fellows frenziedly tried to rig up and get into the water. By the time they did it was all over. As suddenly as it began, the glut ended. Not a fish could be raised. Evidently this phenomenon occurs when certain nymphs maturing at the same time all hatch at once, triggered by the water reaching a certain temperature.

✣ ✣ ✣

>>> *Dry Falls in the Cullasaja Gorge, between Franklin and Highlands.*

Courtesy of the R. Henry Scadin Collection, Ramsey Library, UNCA

Cullasaja Falls,
Highlands, N.C.

Wildcat hunting was a favorite sport with the toughest of the hunters. Two of the best were Zeb Shope and Jess Slagle. They once helped us fight a fire on Roaring Fork back of Silers Bald all one afternoon and until after midnight. They brought their dogs with them. After subduing the fire, we lay out the rest of the night in the old Chestnut Ridge stock-hunters' camp near Nantahala Gap where it was too cold to sleep, even with a fire. At daylight, the pair of hunters urged me to join them in a cat hunt, which I had never done. They sent me to a stand on the top of Silers Bald, and Jess and one of the other boys took the dogs through the Jarrett Creek laurel and roused a cat. They told me afterward that the race passed right by me. But I had gone to sleep when the morning sun thawed me out and didn't hear a sound. The cat had finally gotten away by crawling into a den down in Moore Creek or Factory Branch.

Once while repairing the road from John Slagle's at the foot of the mountain on Wayah Creek to Nantahala Gap in the divide, three of us camped in an old board cabin at the mouth of Fall Branch, bunking on the floor of one room while using the other to store our rations. Every night a skunk visited us to eat an egg or two. When we tried to discourage him by boxing the eggs so that he couldn't get at them, he waked us by rattling our canned goods in his search for food. So we plugged the holes in the storeroom, but he just slipped into our sleeping room and proceeded to walk over us to get to the storeroom.

He certainly wasn't afraid of us. When we shined a light on him rummaging about in the storeroom, he'd peer at us scornfully with his beady little eyes, then go on about his business. Exasperated, we carried in an empty box and let him watch us put an egg in it. When he crawled in after the egg, we slapped a board over the top and took box and skunk out to the old stable and left him locked up until noon the next day, hoping that would cure him, since one of our crew, eighty-year-old Uncle Jake Cloer, wouldn't let us eliminate the animal.

"That's a good cat, boys," Uncle Jake said. "Don't ye hurt 'im. When this job's done, I'll take that cat to the Ol' Woman. She shore likes cats!"

The very next night, though, the skunk was back in our storeroom, as indifferent to us as ever. This was too much even for Uncle Jake, and he agreed that radical measures were necessary.

It was more difficult this time to toll the skunk into the box. With the box open at the top, he wouldn't crawl in, so we turned the box on its side and placed the egg in the back. The skunk went in after it but managed to roll the egg out before we could slip the board over the front. We thought that if we let him eat the egg, he might leave for the night, preventing us from trapping him, so we decided we'd better get the egg away from him, which he was trying to gnaw under a little table.

The sight I'll never forgot was Uncle Jake, in long drawers, down on his hands and knees, trying to frighten that skunk into abandoning the egg. Careful to maintain what he considered a safe distance, Uncle Jake, his disheveled gray hair sticking out in all directions, would rear up on his knees, throw his hands out, and "sh-h, sh-h, sh-h" for all the world like a woman trying to shoo chickens. Mr. Skunk would pause in his gnawing, arch his back, throw his tail in the air menacingly, and fix his beady little eyes on Uncle Jake. The rest of us would back into a far corner, and Jake would retreat momentarily, only to renew the efforts when the animal went back to his egg.

This went on for some time, accompanied by roars of laughter and much free advice to Uncle Jake on our part. As for the skunk, the interruptions in his efforts did succeed in stopping him from cutting through the egg shell, and after a while, the egg rolled away from him a few feet. Uncle Jake beat him to it. Finally, we tolled the skunk into the box and got a board over it. This time, parading down the moonlit road in our underwear, we took box and cat about two hundred yards from camp, placed them against a big log, and slipped a stick of dynamite with lit fuse into the box. Little more than a part of the polecat's tail could be found next morning, but the odor permeated the woods, reaching all the way to camp.

✣ ✣ ✣

One March day in 1924 on a return to my headquarters at Franklin from a field trip, I found a telegram from Washington instructing me to report at once to the headquarters of the Pisgah National Forest in Asheville for temporary assignment. I had heard many favorable things about Verne Rhoades, supervisor of the Pisgah, and was delighted by the opportunity to gain firsthand knowledge of the best known of the southern Appalachian national forests.

My new assignment was as an emergency game warden. At that time, Pisgah was the only Appalachian forest in which the Forest Service attempted to administer a game refuge. A good stock of deer and some bear came with this ninety thousand-acre property when it was acquired from George Vanderbilt's estate, and both deer and bear were multiplying under protection. Few deer or bear were left elsewhere in the mountains at that time (though they have since returned), and poaching on the Pisgah refuge was a real problem. Some of the worst offenders lived on farms adjoining the forest and kept deer dogs. Few wardens were available to protect the game and break up the poaching.

The regulations provided that dogs found on the refuge were to be shot by the wardens. In several instances the regulations had been enforced, and an increasingly bitter feud had developed between the wardens and some of the local citizens. In one instance, one of the worst poachers had run into Supervisor Rhoades and Assistant Supervisor A. C. Shaw in the forest, held them up, and relieved them of their pistols. He told them that some of his kin had lost a dog or two to the guns of the wardens.

In the spring of 1924, the friction finally erupted in the form of a rash of forest fires set simultaneously at widely separated spots. They burned several hundred acres of forestland, almost destroying the ranger station on Bent Creek where Ranger Fink lived. Word went by grapevine from the poachers to the forest officials, "If you kill our dogs, you'll fight fire!"

This was a grave challenge to the Forest Service, for an arsonist can easily burn

Elliot Lyman Fisher—Cataloochee Ranch Collection

This 150-year-old cabin, in the Hominy Creek section of North Carolina, was typical of the homesteads that populated the southern Appalachians throughout most of their history of habitation by white people. In the background is Mount Pisgah.

thousands of acres without being caught. To help cope with the situation, three of us were rushed in to stay with Ranger Fink at Bent Creek for the remaining six weeks of the spring fire season. We patrolled the mountains day and night with loaded rifles, watching for poachers, dogs, and fires. Two were on daytime patrol sleeping at night, the other two on night patrol sleeping by day.

In daytime, from the tops of knobs within the borders of the forest, we could see the homes of some of the poachers, and many a day I lay for hours, rifle beside me, watching for them or their dogs to start toward the forest. To this day, I am not sure what I would have done if I had seen one enter the forest alone or with a dog. On night patrol, one of us would take the top of the ridge on the north side of the Bent Creek watershed and the other the south rim. Usually, I would follow the crest of the ridge until I reached some knob offering a wide view. Then if it were warm enough, I would sit propped against a tree, listening to the soft night sounds as I waited for the music of a deer-chasing dog and watched for the telltale flicker of a tiny flame somewhere off on the dark ridges, the first sign of a forest fire being started. Actually, I dozed about half the time.

Night watch could usually be abandoned by 1:00 or 2:00 a.m., for by that time, dew had dampened the dead leaves and twigs of the forest floor. A firebug would be foolish to try to start a fire then, for either the fire would not burn at all or would burn so slowly as to cause little damage before being spotted and extinguished the next morning.

Not a single unusual event occurred that spring. No poachers, no fires—and I was just as happy. My greatest fright came one night when I was sleeping peacefully in the dim moonlight propped against a tree on a high knob on the south rim of the watershed and was awakened by a skunk innocently scratching the leaves ten feet from me. I awoke with a start, swinging my rifle to ready. Brother skunk bowed up his back in some surprise. But we both held our fire. Finally he sauntered airily off through the leafless brush, and I found it to be time to hike back to the ranger station.

While the four of us were at Bent Creek that spring, a frequent visitor was Watt Hoxed, a mountain man who lived a couple of miles from the ranger station on his own little farm adjoining the forest. From time to time, Watt had worked at odd jobs in the forest. Though neighbor to some of the worst poachers and arsonists, Watt was on our side of the controversy. He was also one of those hilarious human beings so often encountered in these mountains.

Watt was a slender little fellow, perhaps fifty years old, with crackling blue eyes. When he grinned or laughed, humor lines etched all over his face stood out sharply and the corners of his wide mouth turned downward. Mostly Watt told us simple tales about his domestic animals—his wife's chickens, especially strutting roosters, his feist dog, his bull yearling's unsatiated amorous desires. Perry Davis liked to egg him on until all of us were convulsed with laughter.

Watt had a peculiar way of holding his right hand as he wound out his stories, a mannerism as vivid to me today after forty-five years as if I had just seen it. He bent his three middle fingers like a half-clenched fist, pointing the little finger in one direction, maybe slightly crooked like the little finger of a tea-drinking matron, stretching the thumb straight out in the opposite direction just as far as it would go. As he talked, he would slowly wave this hand, held upward from a bent elbow, backward and forward to emphasize certain points. Somehow this silly gesture added to the hilarity of the stories.

✣ ✣ ✣

In the summer of 1925, I accepted a job as logging superintendent of a lumber company deep in the swamps of south Georgia, but it didn't last long. It was a drought year, and sluggish rivers were so low that logging barges couldn't operate. To make matters worse, my mountain fever flared up again with a vengeance. Within a month of taking the job, I got a telegram offering me the post of district forester of the North Carolina State Forest Service, headquartered in Asheville. Again, I eagerly accepted the chance to move back to the mountains, hardly taking time to wind up the details of the Georgia job.

As district forester, I was responsible for forest fire protection in North Carolina's eleven westernmost counties, the rugged mountain counties of the state. Most of the local people were either indifferent or downright opposed to trying to stop woods fires. They had all kinds of reasons for burning the woods—to kill rattlesnakes, bean beetles, or insects; to make grazing easier for livestock and hunting easier for people; or simply to

Jane Alexander

This typical view of the Smokies evokes the original Cherokee name Gad-a-lu-tse, *variously translated as "wave upon wave" or "standing in ranks." White people corrupted the term to "Cataloochee."*

promote the general health. Few people had ever thought of actually trying to grow timber. The virgin forests were almost gone, and the young growth was so small that it was looked upon as brush, a nuisance. Few realized that if protected, it would eventually grow into fine timber.

We had no fire towers and only the merest skeleton of an organization. Total funds for my district amounted to less than $10,000 a year, which had to cover my salary, office, and other expenses and the salaries and expenses of county and district wardens, deputy wardens, and temporary fire fighters. County wardens got thirty cents an hour when fighting fires, making talks in schools, or posting notices; district wardens twenty-five cents an hour; deputy wardens fifteen cents an hour; temporary fire fighters ten cents an hour. Before resorting to paying wardens or fire fighters, we tried to get them to contribute their services.

The summer of 1925 was the driest on record up to that time, with high-altitude hardwood and spruce forests dry as tinder. Making my first round of the district after starting my new job, I reached Bryson City on an August afternoon. There, I learned that a great fire was burning on the lands of the Ravensford Lumber Company in the Balsam Mountains. The fire had already spread over several thousand acres of high mountain land, where the company had logging camps, railroads, and millions of feet of cut logs. When I phoned the company headquarters at Ravensford, the superintendent urged me to come at once, saying the fire which had been burning for two or three days had already caused severe damage. He suspected a disgruntled former employee of starting the blaze.

Just before dark two men from the company office and I set out on a putt-putt motor car on the company railroad. The wife of a logger who was just recovering from an appendicitis operation went along. I held her in my lap on the journey up the rough railway to prevent her stitches from tearing.

On the way up, we came to a high trestle, the end of which was burning briskly, clearly the job of an arsonist. We put out the fire before it could destroy the structure and reached the main logging camp at Round Bottom well after dark. Since it was too late to go on to the fire area six or seven miles beyond, we spent the night there, leaving at four the next morning on a log train that crossed through Pin Oak Gap into the Cataloochee watershed about daylight. The great Cataloochee basin was covered in smoke, and it was impossible to tell where the worst of the fire was centered.

After crossing through Pin Oak Gap into the Cataloochee high country toward Balsam Corner and Mount Sterling, the railroad line bore left under Beech Gap. At the site of the outer logging camp in the spruce belt of the high Smokies, flames from log dumps could still be seen through the smoke below us and above us. Scattered along the smoke-hidden railroad line were skeletons of log cars, skidders, and other machinery, but the main fire had already swept on up the slopes and had died out at the top.

There was nothing further that could be done. The loss from the fire, one of the few to damage standing spruce and balsam timber in the normally wet southern mountains, was later estimated at over $200,000. This was my first trip into the Smoky Mountain high country and a bleak introduction to my new job as fireman for the state.

As the dry year went on, trying to control forest fires with our meager funds and organization became much like trying to sweep back the ocean with a broom. On late fall nights, one could ride along the highways and see the glow of several fires at once over on the ridges. At times, it seemed as though there would be no unburned woodland by the time fire season was over. But in reality, we successfully fought scores of fires. Many others died out of their own accord at streams, roads, or the edges of fields, and still others were put out by private citizens without ever being reported to our office.

Fires in the southern Appalachians aren't normally severe, particularly in the fall. Generally, they burn only leaves on the ground, low underbrush, and downed logs and tree laps. In hardwood areas, fires almost never "crown," so that direct damage to standing timber is usually light. Many woodland owners, not recognizing fire's indirect damage of preventing new timber growth, wanted their lands to burn. Our job included trying to educate the indifferent owners by leaflets, posters, talks in meetings and schools and, where feasible, by prosecutions.

But it was hard to make prosecutions stick, largely because there was little public sentiment in favor of stopping fires. This indifference extended to local prosecuting officers. Unless we had an exceptionally clear-cut case of aggravated incendiarism, we seldom took anyone to court, but simply tried to get the party responsible for the fire to pay the costs of suppression. Even so, we did get some convictions, mostly with light fines of ten dollars or so and court costs of four or five dollars. These were helpful, since word spread rapidly among the culprits and their neighbors, astonishing people who for generations had expected or wanted the woods to burn.

Once in Polk County I saw smoke in a small patch of woods and found a man trying to beat out a fire. He readily admitted that the fire had blown from his burning brush piles. After we had put out the fire, I told him I would have to prosecute him and pulled out my little leaflet explaining the fire laws. He was amazed that there was a fire law, but readily submitted. When I took him in to the Justice of the Peace in Columbus, that official, equally astonished, asked me to handle the matter, since he did not know what it was all about. So I acted as arraigning officer, prosecuting officer, and defending attorney. With the J.P.'s concurrence, I fined the man ten dollars and two dollars in costs, the latter representing the gas and oil for hauling him into town and back home. I wrote out the man's check for him, drove him home, and was invited to supper.

All cases were not so simple. On one occasion a prominent surveyor in Jackson County who was also something of a power in local politics, lost a chain pin in the leaves while surveying. He set the woods afire trying to find the pin. An Indian warden named Sibbald Smith put out the small fire, secured a warrant for the man, and called me to meet him for trial at the local J.P.'s home in Whittier. The accused surveyor began the proceedings by abusing the Indian warden and demanding to know who was going to prosecute him. Angered by his attitude, I said I was, but the surveyor's political standing frightened the little J.P., who refused to rule in the case. So I took the warrant to the solicitor of the state superior court in Waynesville, who promised to do something about it. But needing the surveyor's support in his next race for office, the solicitor never called the case for trial.

In those days we had a law which gave us the right to summons anyone to help suppress fires. The Transylvania County warden, Elon Galloway, had been plagued with so many fires that he was about to give up in fatigue and frustration. So one night, Assistant State Forester W. C. McCormick and I went to help fight a fire burning briskly only two miles from Galloway's home in Rosman. Galloway told us about three mountain boys courting three sisters living nearby, so we decided to summons them to help fight the fire. Knowing the boys would be reluctant, especially if they were in their "courting clothes," we carefully planned our approach. While Galloway and McCormick slipped around to the back door, I knocked on the front door, telling the girl who answered that I would like the boys to help put out a fire. I could hear sudden scuffling in the back room as the boys made a break for the back door. As they burst forth, Galloway and Mac caught them one by one. The boys meekly went along with us to fight the fire—which we put out in about an hour—then returned, in good spirits, to the serious business of courting.

Galloway told us that a huge woods fire was sweeping up toward the North Carolina line from South Carolina in the Horsepasture and Whitewater river country below the Blue Ridge and beyond an isolated community called Bohenie. Leaving Galloway at home to look after the rest of the county, Mac and I went out the next day, arriving that evening at the home of Deputy Warden McCall in Bohenie. The presence of strangers frightened the McCall infant into a tantrum, causing Mrs. McCall to interrupt her supper preparations long enough to reach into the side cupboard and bring out a quart fruit jar of white corn liquor. She fed about half a tablespoon to the tiny baby, who in just a few moments was sleeping peacefully in my arms. Soon we enjoyed a good supper of hog meat, hot biscuits, potatoes, and coffee. Mac and I slept that night in a double bed in the attic, and nary a sound was heard from the infant, who awoke next morning chipper and bright.

In some counties, it was nearly impossible to persuade a good citizen to accept the job of warden: the pay was poor, the work difficult, the areas isolated, and the roads almost impassible. But mountain newspapers were helpful in educating the public about fire protection, carrying articles whenever we would supply them. Perhaps the greatest influence in changing people's attitudes was the example and word-of-mouth work of a few native men who took up the battle early and worked for the cause, often in the face of ridicule by their neighbors. Some individuals who stood out in those early discouraging days were Walt Jenkins of Swain County, George E. Brown of Haywood County, Forest Denton of Graham County, and Elon Galloway of Transylvania.

Walt Jenkins had an especially tough county. Much of Swain was in timber then as it is now, with several big lumbering operations covered with half-rotted tree laps and slashings, making fire fighting particularly difficult.

One April afternoon Walt called to tell me that he and a crew had been fighting a fire for two days on the divide between Noland and Forney creeks. His crews were worn out, but he had a few men watching the lines who would stay all night, even though completely out of food. I loaded my coupe with canned goods, bread, and chocolate bars, and started out in midafternoon. Walt lived on Noland Creek about six miles up the creek from the Tuckasegee River, and the road I had to take from Bryson City was under construction and impassible in spots. At one point, I had to drive my car up on the Murphy Branch railroad tracks and negotiate a short trestle to reach Walt's house before dark. We went on to the fire that night, and by backfiring along some old skid roads brought the fire under control by daylight.

North Carolina was one of the first southern states to undertake organized state forest fire control, and J. S. Holmes—a modest, charming, dedicated gentleman—was the state forester. To know him was to love him, and I consider it one of the great privileges of my life to have been able to work with him. Even then his health was not good, and during the winter of 1925-6, he had to take a month's leave of absence, leaving me in charge of the Raleigh office. While the experience was interesting, I got another bad case of mountain fever in my brief absence from the hills.

✣ ✣ ✣

The tragic Florida real estate boom of the twenties overflowed into the Western North Carolina mountains in 1925. High-pressure real estate offices opened all along the main streets of Asheville, and farmlands near the city and further out were bought up and

George Masa—Courtesy of the N.C. Division of Archives and History

Pack Square in Asheville at about the time the Western North Carolina real-estate boom was collapsing.

subdivided—enough lots to furnish one apiece to a large part of the United States' population. Water mains and electric lines ran far out into the country. Staid and substantial businessmen, abetted by banks, speculated feverishly, buying a lot one day with the paper profits from an earlier sale, selling the lot the next day for an enormous cashless "profit," in a tottering frenzy of greed that was ultimately doomed.

Although millions of acres of wild mountain land exist in Western North Carolina, every tract seemed ripe for speculation. Before the fever ran its course, great blocks of cutover acreage changed hands, some sold by absentee lumber companies smart enough to unload otherwise practically worthless lands at exorbitant prices. After it became clear that these tracts could not be subdivided and sold to suckers, some were converted by their new owners into hunting or fishing clubs. Other owners, however, constructed expensive developments complete with artificial lakes, graded roads, clubhouses, golf courses, swimming pools, and cottages. Only a few of these survive intact today. What one sees instead are decaying and dilapidated cottages slowly rotting back into the earth, golf courses reverted to cow pastures and swimming pools to frog marshes. It was fun while it lasted, but untold millions of dollars were lost, and the financial backwash almost destroyed Asheville, Buncombe County, and other mountain towns and counties.

In 1926, I was hired as forester to look after the six thousand acres of land belonging to the Chimney Rock Corporation in the Rocky Broad River gorge and its tributaries. Chimney Rock was perhaps the most elaborately planned of the developments, and millions of dollars went into the construction of a dam, golf course, roads and streets, hotel, administration buildings, water and sewer systems, and power plants.

Eager to do something to justify my salary, I hired a crew to build a combination riding trail and fire-break as close to the border of the property as grades would permit. We had hardly gotten started before the development's board of directors shut off our money. The mountain region's real estate boom was already beginning to collapse, and Chimney Rock Mountains, Inc., was feeling the effect. To keep me busy, my employers assigned me a small crew of Negro men to trim underbrush and thin young timber along the roads and on unsold lots. Before long, even this activity was curtailed. With nothing to do, I wanted to resign, but Dr. Morse, president of the development, asked me to stay on, and I agreed.

During the summer, bus load after bus load of "prospects" were hauled in from all over the Southeast to be shown the beautiful properties and high-pressured into buying one or more lots, which ranged in price from about $1,000 to $10,000. Some of these people may have been legitimate prospects, but most were sightseers, school teachers, stenographers, clerks, and so forth, who could not possibly finance a lot and house. Part of my duties was to dress up neatly and help entertain the visitors. This was not forestry, but it was not unpleasant either. So I hung around all summer, taking girls on hikes to Chimney Rock, to the caves up on the face of Rumbling Bald Mountain, over the golf course under construction, or to the properties being developed. But I could feel the chill winds of imminent failure, for few lots were being sold.

I had time to make many side excursions during those Chimney Rock days. Thomas Dixon was developing his Artists' and Writers' Colony near Mount Mitchell. Another group was spending a million dollars on Hogback Mountain near Tryon. Lake Tahoma near Marion was being built and the surrounding lands cut up into lots, and innumerable smaller residential subdivisions were under construction near mountain towns and cities. I also visited the older mountain resorts—Blowing Rock, Linville, and Highlands. To me, Linville was the most attractive, with its levelness, its luxuriant forest and floral surroundings, and its charming bark-covered hotels and summer homes. Chestnut bark was still to be had, and a small shop at Linville made chestnut bark shingles. Properly applied, these shingles give a most attractive, rustic appearance to cottages that holds up well. Poplar bark is just about as good and lynn or basswood is sometimes used, but none was as readily obtainable as chestnut.

By fall, the squeeze came. The sales organization had largely faded away, the buses had been discontinued, and many of the engineering projects, such as street grading and paving were sharply curtailed. The fall fire season was approaching. In November, I

Herbert W. Pelton—Courtesy of the Ewart M. Ball Collection, Ramsey Library, UNCA

A top scenic attraction since the nineteenth century, Chimney Rock gave young forester Tom Alexander a first-hand taste of the land boom and collapse that shook Western North Carolina to its roots in the late twenties. Hired to manage property being promoted as a real estate development—like

Herbert W. Pelton—Courtesy of the Ewart M. Ball Collection, Ramsey Library, UNCA

thousands of other properties in the mountains—he watched as buyers lost interest, speculators went broke, and he lost his job. Most of the Chimney Rock property is now a privately owned park, open to visitors. These pictures show the rock itself as it looked during the boom.

Tom and Judy at their wedding in June 1929. The ceremony was held in the garden of the Barksdale home in Malvern Hills, near Asheville.

moved up to the Cliff Dwellers Inn, where I could look out over most of the property for forest fires. I stayed there alone for several weeks until the whole project was closed down for good. The boom was over and so was my job.

I was finally offered a timber-cruising job in south Alabama by the Asheville office of the James D. Lacey Company, a concern that specialized in forest consulting, timber estimating, and appraisal all over the world. After working several months in Alabama, I returned to Asheville and hung out at the Lacey office hoping for another project.

Lacey's Asheville representative, Robert R. Hope, hailed from Brooklyn and found it difficult to get along with southerners, especially the timber men he dealt with. I accompanied him to several Appalachian timber tracts Lacey represented in Kentucky, Tennessee, North and South Carolina, and Georgia. I also learned the office end of Hope's responsibilities. After I had been there about three months, Hope made a trip to Brooklyn and never returned, and I inherited his job by default. It suited me well, providing me an acquaintance with many timberlands and lumber companies.

I also got to know many of the younger businessmen of Asheville. One group owned a hunting and fishing lodge, actually a former mountain farm that had been incorporated into Pisgah National Forest. I was invited to join the group's members, including wives and girlfriends, on a four-day house party and pack trip on horseback. After spending the first night at the lodge, we rode into the forest and camped out for the next two nights, sleeping around a big campfire, pretty much like hibernating snakes.

One member of the party was a golden-haired, blue-eyed girl of eighteen who immediately caught my fancy, both for her physical attractions and the spirit she showed in volunteering for the cooking, dishwashing, and clean-up chores at the lodge and in the wilderness. It was also clear that she was athletically inclined. As acquaintanceship ripened into friendship, I found her to be as well informed in the arts and literature as in the natural world. Her name was Judith Barksdale, and she was the daughter of Dr. George Barksdale, an Army surgeon assigned to the Oteen military hospital outside of Asheville.

In a courtship of about three years, I overcame the various obstacles to marriage, including persuading her and her parents and buttressing my financial future. During this period, Judy and I gratified our mutual love of hiking mountain trails and driving backcountry roads. We were finally married in June of 1929 in the flower garden of her parents' home near Asheville. After a wedding trip to Nova Scotia, we returned to a rented house in Asheville.

Courtesy of the Ewart M. Ball Collection, Ramsey Library, UNCA

The heavy rainfall characteristic of the region pretty much guarantees that mountain travelers are seldom far from a source of water.

Heart of the Mountains

The movement to establish a national park in the southern Appalachians, which had been simmering ever since the 1890s, came to a head in 1927 with the passage of a bill by Congress authorizing the federal government to accept donated lands in the Great Smokies. The park was to lie along the boundary between North Carolina and Tennessee, roughly half in each state, on lands belonging to lumber companies, estates, hunting and fishing clubs, and thousands of small mountain farmers and stock raisers.

Since Congress had decreed that no federal funds were to be used to acquire the lands, the two states involved instituted a drive for subscriptions, but it soon became evident that sufficient funds (an estimated $10 million) could never be raised that way. John D. Rockefeller, Jr., offered to match all funds dollar for dollar up to $5 million as a memorial to his mother, Laura Spelman Rockefeller. In addition, the states voted bond issues. In combination, these provided the necessary amount.

The original plan had been to purchase the land by direct negotiation if possible, but the state commissions were also given the power to condemn if necessary. The two state commissions organized appraisal crews to set values on the various holdings, but most of the large-tract owners proved unwilling to accept the state valuations and arranged to have their own appraisals made. Among those hired to do the appraising was my employer, the James D. Lacey Company. In doing so, we first had to estimate the amount of merchantable timber—to "cruise" the lands—and then to set a unit value on both timber and land.

One of our clients was the Ravensford Lumber Company, which owned some thirty-five thousand acres of fine virgin timber on the Raven Fork of the Oconaluftee River in the heart of the Great Smokies. We set up a tent camp in the remote Three Forks basin, north of Qualla Boundary, the Cherokee Indian reservation.

The name Three Forks usually designates a twenty-thousand-acre forested bowl suspended high in the angle between the main Great Smoky range and rugged Balsam range. But the name derives from an incomparable gem of sylvan beauty lying at the center of the basin. There, deep in virgin timber, three streams flow together in a crystal pool some fifty or sixty feet wide and ten to twelve feet deep. The water of this pool is so clear that the deepest pebble seems to lie near the surface, which mirrors tall dark columns of spruce and hemlock. Brook and rainbow trout, large and small, hang and waver in the depths, so well nourished that they rarely show much interest in a fisherman's lure.

George Masa—Cataloochee Ranch Collection

Cattlemen and hog-rangers constructed dozens of primitive communal shelters throughout the Smoky wilderness. George Masa photographed one such cabin near the fishing camp at Three Forks.

George Masa—Cataloochee Ranch Collection

In this 1931 picture are, from left to right, Bill Davis, Lucy Hildebrand, Barbara Ambler, Jewel King, Walter McGuire, Judy Alexander, Bill Shuford, and Tom Alexander, all of Asheville.

In those days, a picturesque little stock-hunters' log cabin stood in a clearing a few steps from the pool. The roof was of hand-split balsam shingles and the chimney mud-chinked. Slightly larger than a dog kennel, it was the center of a superb campground. The little clearing in which it stood had been gradually expanded over the years by hunters and fishermen cutting trees for firewood. Horses had grazed here, keeping down the briars, while their owners fished or hunted. The horses seemed to have found wildflowers distasteful, for these flourished in amazing variety and profusion. Horsemint, orchids, coneflowers, and hosts of others—their reds, purples, and yellows vying with the pinks and whites of the rhododendrons and laurels of the surrounding forest.

George Masa–Cataloochee Ranch Collection

Passing through the tunnel-like gorge a few miles below Three Forks, Raven Fork becomes a kind of natural millrace.

>>> Taken by George Masa during one of his visits to the Three Forks fishing camp, this photograph captures the stately lushness of the virgin hemlock, spruce, fir, and rhododendron forests of the higher Smokies. All but consumed in the Appalachian logging boom in the first third of the twentieth century, only patches of these virgin conifers remain. Later, the chestnut blight wiped out the region's dominant hardwood species. The detritus from these events helped fuel the devastating forest fires that swept over the mountains for years afterward. More recently, another mysterious stress, believed to be a combination of air pollution and insect infestation, has decimated the beautiful "she-balsam" (Fraser fir) forests that once blanketed the high country.

George Masa–Cataloochee Ranch Collection

For seven miles above the pool and clearing, right up to the crest of the ranges forming the bowl's rim, the forest is unbroken and the laurel and rhododendron understory virtually impenetrable. Below, Raven Fork meanders quietly for several miles along an almost flat, shadowed basin floor. Then, five miles below the campground, the ridges forming the rim press together. Seeking an outlet between them, the water takes a twisting, swifter course as if to escape from the bowl before the ridges come entirely together and imprison it forever. It enters a gorge, tunnel-like because of the dense, overarching forest. There, the streambed is choked with huge moss-covered boulders, many larger than houses. Among these, the water has scooped eerie, seemingly bottomless, blue-green pools.

After a few miles more, the stream's frantic dash through this black canyon abruptly ends. The ridges spread away, the forest gives way to clearing and, suddenly, the water gently ripples across the flat, open floor of the Cherokee reservation, laughing in the sunlight, happy to be out of the wilderness.

The first trail into Three Forks twisted up through the gorge from the Indian lands. Over it were brought, at great effort, the rainbow trout fingerlings from which the plentiful population of the river has descended, partly displacing the native speckled trout. Later, a dim trail was worked out from Straight Fork, over Hyatt Bald on the south rim and thence out Breakneck Ridge and down into the clearing. That horses negotiated both of these trails at one time is evidenced by fragments of horseshoes found along the way.

When the timber in the bowl was to be appraised for the park, another rough and steep horse trail was blocked in from Smokemont, the nearest white settlement, nine miles away through the forest and over two ridges. Over this trail the supplies for our timber cruisers were packed in. Later another dim trail was run up Middle Ridge from the Three Forks clearing to the Balsam rim and thence along this and the main Smoky range to Mount Guyot. The cruisers used this trail as a baseline while estimating the timber in the bowl. Hikers later used it to ascend Mount Guyot from their camp on the stream.

When our timber-cruising was done, the lumber company that owned the land rented our camp and equipment for the use of crews surveying logging railroad lines through the tract. While these railroad crews were at work, the Great Depression hit. The Lacey company went broke in late 1929, owing me some $600 of back salary I hadn't collected while off in the woods. The company was put in the hands of a receiver, who informed me I would not be paid. At about the same time, the bank in which I had some money deposited closed, so like many of my friends, I was caught with neither money nor job. Not long after that, Judy and I became aware that a child was on the way.

✣ ✣ ✣

One afternoon a friend, Roy McDuffie, who had gone broke in his business of making concrete tiles for sidewalks, came calling and, in the course of our conversation about hard times, suggested that perhaps we could somehow get into tourism. That suggestion brought to my mind the company equipment that was still up at Three Forks.

I immediately wrote the receiver of the James D. Lacey Company, suggesting that I

George Masa—Cataloochee Ranch Collection

The former timber-cruisers' camp at the Three Forks pool made a tranquil if primitive summertime trout-fishing resort for young Asheville businessmen willing to hike or ride the nine miles over rough terrain to get there. Fish were always plentiful.

would take that camping equipment in lieu of the salary owed me. He agreed, and I suddenly found myself in the tourist business, the owner of an outfit of tents, cots, mattresses, and cooking equipment, already in place beside one of the finest trout streams in the southern Appalachians even if only the most dedicated fishermen could actually get to it. Roy and I knew there must be thousands of potential customers who would want to fish that stream. And there were probably plenty of nonfishermen who might want to see beautiful country and magnificent timber.

I inquired when the lumber company expected to complete the railroad surveys so that we could take possession of the camp, then lined up our first fishing party. I drove that group, twelve to fourteen men and women, to Smokemont and led them afoot the nine tough miles into camp.

When we got to Three Forks, we found the camp still occupied by the surveyors, who had not finished their work. Instead of having cots to sleep on that night, we settled down around a campfire. Though it was not an altogether comfortable night, those first customers were fine sports. We had packed in some food and had caught a few trout, so we had enough to eat for supper and breakfast. In a few days, the surveyors left, and we got things better organized.

I had hired an experienced woods cook, Avery Gouge (pronounced "Googe"), who had cooked for timber-cruising parties throughout the Smokies. Avery stayed at Three Forks between my parties, spending his time making the camp more attractive. Along with all the fish he could eat, Avery had other food that we packed in for him, so aside from loneliness, which didn't worry him much, I think he enjoyed himself.

Roy McDuffie lined up customers from Asheville and I led them into camp. If they had any trout-fishing skill at all, we could practically guarantee them all the fish they

George Masa—Cataloochee Ranch Collection

The exigencies of the Great Depression forced Tom and Judy Alexander into the hospitality business in 1929. Here, seated under the tent flies of his former employer, the bankrupt James D. Lacey Company, Tom (right) dines with some of his trout-fishing guests at the Three Forks camp. Cook Avery Gouge stands at left.

wanted. But to make sure we would always have some to eat for that first meal, Avery fixed up a small pool in the woods about a hundred feet from the tents where a small spring rose. He kept it stocked with thirty or forty trout.

Soon familiar with every pool, rock, and waterfall of the Raven Fork stream and its tributaries, I could advise guests where to go and the best way to fish. Above the big pool, the left-hand stream, though small, was alive with speckled trout, with an occasional rainbow to be found in the larger pools.

It is about five miles through wonderful country to the headwater spring of this stream. The giant timber, mostly spruce and balsam, and the unbroken, almost impenetrable undergrowth of rhododendron and laurel creates a jungle through which comparatively few white people have traveled. The streambed, the only route, is choked in many places with treetops and broken bushes that must be crawled through or over. Waterfalls and rock ledges abound. Soft and springy moss grows in the water and, together with the rocks and ledges, provides footing up the stream.

Early in this century fierce rainbow trout brought in from the west pretty much displaced the smaller eastern brook or speckled trout in many of the larger streams. But the brookies were still plentiful in streams above the Three Forks camp. In this 1932 photograph, a good day's catch of both rainbows and speckleds dangles from a string.

George Masa—Cataloochee Ranch Collection

Hiker Jewel King examines a freshly bear-clawed tree on Mount Guyot in 1931.

George Masa—Courtesy of Barbara Ambler Thorne

This stream had been fished by Indians before the first whites ever came, but the native brook trout were still plentiful. Most of these beautiful little rose-spotted orange, white, and black fish were small—mostly four, five, and six inches long, though an occasional one up to twelve inches could be taken in the larger pools. Rarely was a rainbow trout found up this high. It was not uncommon to find the tracks of a bear on the sandbars along the stream.

Most likely, hunting and fishing were the principal motives behind the location of the stock-hunters' cabin at Three Forks, stock hunting being the pretext that menfolk offered womenfolk for getting away from home. Along with fish and bear and turkeys, men would have been interested in still hunting for "boomers," the little red squirrels that are about half the size of the common gray squirrel. Boomers figured as ingredients of both tasty gravies and folk tales, being legendary for their speed and cockiness much like Brer Rabbit in Negro tales of the flat country. It is said that when harvesting acorns for winter storage, a boomer will cut an acorn off and then run down the tree and dig a hole in the ground for it to fall into. According to one old story, a boomer was feeding atop a tall hickory-nut tree as a lightning storm comes up. He spots a bolt of lightning headed his way, runs down the tree, stands off to the side, and hollers to the oncoming lightning, "Now, split it!"

✣ ✣ ✣

Another favorite quarry was the ruffed grouse, or "pheasant," as the natives call it. These were often heard drumming in the woods surrounding the camp. One morning, I was leading in a horseback group through a stretch of hardwoods. My old mare suddenly spread-eagled and refused to move further. I heard fluttering and hissing noises beneath her head and, looking down, discovered that a mother grouse was attacking the mare. Her neck feathers stood straight out, her tail was spread and dragging, and she was jumping knee-high up the mare's front legs, clawing, pecking, and slapping with her wings. Wise from years in the woods, my steed quickly recovered her equilibrium and stood there taking the beating.

I knew what it meant and looked about in the nearby greenery for the babies. I quickly spotted four or five struggling through the twigs and stems of the forest floor, just above the upper trail bank, and pointed them out to the riders behind me. The little brownish fluffy chicks, evidently newly hatched and hardly as big as ping-pong balls, were hunting leaf shelter while their mother held our whole string of horses at bay. Even while we watched, the little fellows disappeared completely. Each had simply melted into last year's leaves under one of the mayapple or loosestrife plants that covered the forest floor.

The ruffed grouse is wise and wary in hunting season, yet almost indifferent to humans at other times. This unpredictable thunderbolt of the mountains has won the affection of all who know it. It has never been "on relief" as have quail, wild turkey, and ringnecks, but has survived and thrived by its own clever devices.

In North Carolina, grouse are confined exclusively to the mountain regions. Though primarily forest birds, they also like the edges of old clearings that have grown up in saw briars, grapes, sumacs, and haw bushes. When especially good food is available, as many as ten or twelve birds might gather on occasion, but more typically, they feed in small groups of three or four.

They are perhaps this country's most omnivorous game bird. In analyzing their crops and gizzards, investigators have found buds, twigs, fruits, berries, seeds, grains, acorns, flowers, leaves, and roots; eggs, larvae, cocoons, pupae, nymphs, and spiders; snakes, small birds, clams, mussels, and cottontail rabbit fragments; hair, bits of glass, lead shot, and roofing nails. In winter, grouse commonly eat the green leaves of mountain laurel which is poisonous to domestic animals. Dogs and cats are sometimes mildly poisoned from eating the offal of grouse that had been dieting on laurel leaves, and I myself have felt some ill effects after eating laurel-fed grouse.

The male grouse is a ventriloquist. In spring and summer and on clear warm days in the fall, he gets up on a log and beats the air with his wings, making a drumming noise that can be heard for half a mile. The beat starts slowly but builds to a crescendo: *thump, . . . thump, . . .thump, thump, thump, thu, thu, thu, thr, thr-r-r-r-r-r-r-r.* You think the sound is coming from a long way off and you start toward it.

Next time the grouse drums you realize you have started in the wrong direction, and you change course. You ease up through the woods toward where you think he is, moving only when he is drumming, since the cock grouse presumably is not alert to anything going on about him then. Thinking you need to make quite a few forward spurts to reach him, you move three or four times, but suddenly the drumming stops for good. Old brother grouse was closer than you thought, saw you coming that last time, and has glided quietly off the log and down through the thicket. (He can fly quietly when he needs to.)

With experience, you may actually slip up on him while he is drumming. One hunter I know managed that but didn't realize how close he was until he stepped around a clump of bushes and found himself face to face with the grouse standing alertly on a log less than fifteen feet away. Each stood motionless, the bird obviously planning his escape, the hunter trying to figure how he could bring his gun up for the kill. Since there were no bushes or trees for the bird to dart behind on his takeoff, the hunter "knowed he had him." But the old grouse simply dived off the log, and flying not a foot off the ground, skimmed out of sight into the brush fifty feet away, while the outwitted hunter blasted a hole in the top of the log.

Once, my little feist dog flushed a grouse in the open on a steep face of a bald near a spring. As grouse often do, this one flew up into the only sapling nearby and sat in plain view on a high limb—entirely too far for me to kill him with No. 7½ shot. Hoping to be lucky enough to stun him, however, I fired both barrels in quick succession. Instead of flying off as any sensible bird should, this one hopped to another limb on the other side of the trunk. The tree was only about six inches in diameter, and a grouse is about seventeen inches long, yet this bird was completely invisible. I reloaded my gun and moved cautiously forward, sure I could get him if he sat there long enough for me to get in gun range. He let me get within twenty-five feet of the tree. Then suddenly I saw the tip of his tail swish upward behind the tree trunk, and in the same instant, that bird plummeted straight down to the ground and skimmed at unbelievable speed down that steep slope, so close to the ground that little undulations in the grass hid him from sight half the time as he headed for the woods nearly a hundred yards away. I wasted two more shots without touching a feather and went home talking to myself.

Experts claim that the speed of a grouse in level flight is between forty and fifty miles an hour. This isn't as important as the fact that the bird reaches maximum speed in such a short distance. I personally think it starts from scratch at that speed, and if flying downhill, I'm sure grouse can attain a hundred miles an hour in the open. Even in the woods it travels at remarkable speeds, while weaving in and out through the tree trunks.

When motoring along a mountain road, it is not unusual to come upon two or three grouse standing on the edge of the road—usually a cock and a couple of hens, in a group of three. Stop the car, leave the motor running, but don't get out, and likely as not, they won't fly, or at least not for several minutes, but stand there looking at the automobile. After you have watched them for a few moments, one of the hens will begin to ease off toward the weeds on the road bank and finally disappear, followed later by the other two. The last one may fly off after the other two have walked out of sight.

In spring, you'll often come upon a mother grouse and her brood of day-old chicks walking on a mountain trail. The female will start fluttering down the trail ahead of you, for all the world like a badly wounded bird with a broken leg and wing. If you are not wise to her ways, she will lead you seventy-five or a hundred feet down the trail before turning off into the brush. When you've gotten thoroughly tangled in doghobble or huckleberry, she will nonchalantly fly off. If you are wise, you won't follow that deceitful old bird, but will stop and watch the babies scatter and hide, squatting under leaves. Soon after you have moved along, mother grouse will be back where you first saw her, calling her family together with a low humming sound.

While papa grouse is the ventriloquist in the family, both mama and papa can do a superb Houdini disappearing act. As you watch a grouse's bullet-like flight through the

W.C.D. Photo by Reese—Cataloochee Ranch Collection

A male ruffed grouse, caught in his ventriloquy act by a wily photographer.

woods, you are sure you see where the bird landed. Gun ready, you sneak up to the spot you saw him light, but he gets up a hundred feet ahead of you. Another time, he stops short, and you walk right by, almost stepping on him. As soon as your back is turned, off he goes, back in the direction he came from.

I am convinced, as are many others, that the grouse knows just what a gun is and how far it will shoot, even if he is a this-year's bird, never before shot at. If you don't have a gun, you can often saunter toward a grouse standing indifferently in a trail and get within maybe twenty-five feet of him, then watch it walk off through the bushes. But have a gun in your hand and he flies off before you are anywhere within gunshot.

Maybe he is feeding in some bushes in a sort of open place and neither of you sees the other until you are perhaps five feet away. He knows that if he gets up, you will have a good shot at him, so instead of hastily rising, he freezes there in the protective coloring until you have passed beyond him. Then pow! he's off with a roar directly to your rear and out of sight before you can swing around to shoot. Oftentimes, a grouse will sit quietly as long as you keep moving, even if you pass close to him, but if you stop for any reason, he scents danger and takes off, giving you a terrible start.

Grouse used to be far more plentiful than they are now, for they were trapped, lured, snared, and shot by the thousands, for sale in New York and New England. The price ranged from a low of eight cents apiece in Maine one year to seven dollars a pair in Washington, D.C., in 1918. Some of the larger New York and Boston hotels retained their own grouse hunters, and the birds were taken in all seasons, some hunters turning in a thousand grouse a year. Because of the damage they did to apple orchards, some New England states even placed a bounty on them. Now, thank goodness, grouse are classed as game birds, their sale prohibited.

✣ ✣ ✣

Our favorite pet at Three Forks was another bird, much tinier than the grouse and practically indifferent to humans. This was the red crossbill, which normally frequents the spruce woods of New England and Canada. From their avid love of salt, they are called "salt birds" by the mountain people. The early settlers became acquainted with them when they took salt to their free-ranging livestock in the mountains and saw the crossbills dart in to lick logs and rocks. The intelligent little birds had learned the echoing calls of the stockmen and would flash from the forests in among the jostling beasts for their share of the delectable salt.

They are handsome little creatures about the size of bluebirds; males are brick red with carmine rump and dusky, grayish wings; females grayish olive with yellowish rump and breast. The birds' beaks are so sharply crossed as to look deformed—the tip of the lower mandible turning upward and that of the upper downward. These queer bills are shaped for procuring their principal food, seeds from the tight cones of coniferous trees. To pick up salt grains, the birds must turn their heads sideways and thrust their tongues through the sides of their beaks.

A party of riders and I were packing up to leave Tricorner shelter cabin on the North Carolina-Tennessee state line one glorious crisp October morning when a flock of the little birds, excitedly chipping, dashed out of the surrounding spruces and swooped to light on the low roof of the nearby cabin. They squatted there momentarily, then plumped to the ground, hopping about almost on our feet. Just as suddenly they were off again, some to the surrounding trees and the others back to the roof.

We hastily sprinkled the crumbs from a leftover loaf of bread on top of the cabin, but our visitors ignored our offer. After another brief period of excited chipping and fluttering back and forth between cabin and trees, all at once the flock was back on the ground at our feet. Soon we were surrounded by 150 or so, some in the trees, some on the cabin, and the rest hopping about excitedly on the ground where we had spilled some salt.

The little birds seem to have no fear of humans. During the long days at Three Forks, Avery Gouge found them good company. He fed them bread crumbs and grains of salt in the warm oven of our camp stove. The salt birds seemed to like the warmth of the oven and eventually took virtual possession of it. They also liked to join us for dinner, fluttering in to pick up grains of salt scattered on the large plank that served as our table.

✣ ✣ ✣

The original trail to the camp from the south climbed three miles up the timbered face of Hyatt Ridge to McGee Springs, then followed the ridge northward to the end. There it pitched off precipitously down to the big Three Forks pool. It was a tough climb up from the pool, and Avery and I decided that we should cut a better trail.

It was quite a chore to chop out the new trail through the dense rhododendron, but it gave Avery something to do on the days in camp alone. We found many old downed tree trunks bedded down in the rhododendron jungle, too large to chop out with an axe—the only tool we had to work with. We simply cut steps up the sides of the logs to make it easier for hikers to get over.

Rough as it was, this new trail was a great improvement. It led to the hardwood forest along Hyatt Ridge, where there was much more game than in the conifers lower down. Squirrels, grouse, raccoons, and bear were favorite quarry when we tired of fishing. A flock of wild turkeys lived in the vicinity of McGee Springs, feeding on beechnuts and acorns. Occasionally, this flock would take off and fly great distances across Straight Fork Gorge to feed on Ledge Bald, a prominent grass-covered peak where grasshoppers and crickets were plentiful and beech and oak trees produced plenty of nut and acorn mast.

Avery spent much of his lonesome time splitting stove wood and stacking it. His neat rick of wood was always a beautiful sight when we arrived in camp after the nine-mile hike up from Smokemont.

In planning the fishing of the stream, I would start out after breakfast with from two to eight fishermen and go downstream, scattering them out at half- to one-mile intervals, so that each fisherman had some good water in his section. Then I would continue on for perhaps a mile and start fishing back up the stream, collecting my fishermen on the way.

During the two summers we ran the camp, I took in many novice fishermen, but none was ever skunked. With pointers on what kind of lure to use and how to handle it, everyone caught fish. On her first trip into Three Forks, her first trout-fishing attempt, my wife Judy caught nine, an experience that almost made her an addict. She never again objected to my going fishing.

Soon after we started the Three Forks camp, I grew acquainted with a farmer near Smokemont who had three horses and arranged to rent the animals for guests who weren't up to hiking in. The price was four dollars per horse and included a man to accompany riders into camp and later lead them out. I tried to encourage any guest whose stamina I doubted to rent a horse.

George Masa—Cataloochee Ranch Collection

Judy Alexander at Three Forks in 1932, with her catch of rainbows.

In Asheville there lived a noted photographer, a Japanese native named George Masa, who had fallen in love with the southern Appalachians. He hiked out into the mountains at every opportunity, lugging huge cameras. George and I became close friends and spent many happy hours together in the woods. He frequently used our Three Forks camp as a base from which to hike up the surrounding ridges and peaks, taking his stunning pictures. He had also constructed a device to measure mountain miles—the front wheel and handlebars of a bicycle with an odometer attached. His trail measurements, made as a hobby, were used as gospel by many map makers of the era.

Once a young hotel clerk in Asheville, after weeks of urging, persuaded me to take him into Three Forks so he could see the heart of the Smokies. I tried to prevail on him to hire a horse. "How far did you say it was?" he asked.

"Nine miles."

"Aw, that's no tough hike," he scoffed. "When I was a Boy Scout, we hiked ten miles all the time and even took fourteen-mile hikes now and then."

I tried to tell him that Smoky Mountain miles are different from the Boy Scout kind but could not dissuade him. So we started out from Smokemont, each of us carrying a pack of food. It was a hot, muggy summer day, and a recent heavy shower had left the

Hugo Strongmiller—Courtesy of M. R. Gooch and George Ellison

With his massive camera and tripod, noted Appalachian photographer George Masa would wait patiently on remote mountaintops for just the right light.

Courtesy of Barbara Ambler Thorne

trail slick. It was tough going, and shortly after we started, my companion wondered, "How in the world does a horse get over such a trail as this?"

After that first three-mile climb up the open, hot slope, the trail crossed the first ridge and turned down into huge virgin timber. There the path had been washed out by the heavy rains. In many places it was nothing more than a trench full of tree roots and big rocks. At the bottom of the ridge, the trail plunged abruptly into a mountain stream, and for half a mile the bed of the stream served as trail, with rocks and waterfalls as handicaps.

Leaving the stream, the trail climbed steeply up the face of the second ridge and down again into a flat hollow through a beautiful stand of hemlock. Under the dense timber, the sun rarely penetrated, and there the route was a soupy bog of black mud, with accumulated needles and duff of many years. We had the choice of wading through the mud or trying to avoid it by hanging to the laurel bushes that lined each bank, attempting to stay on the narrow ledge between mud and bushes.

In one place we rested, a stretch of the mud was visible for about 150 yards ahead. My worn-out companion had just finished wondering for the dozenth time how in hell a horse got over such a trail, when he looked up and noticed a man coming down the trail. It was George Masa coming down from Three Forks, a red bandana outlining his dark face, pushing his trail-measuring rig before him. The hotel clerk's mouth dropped open, his eyes widened. "My God," he gasped. "Yonder comes an Indian riding a bicycle!"

In this 1931 George Masa photograph, a group of Asheville fishermen and -women huddle about the fire in the Three Forks camp. Visible in left center is Masa's mileage-measuring contraption. Standing at rear is cook Avery Gouge and going clockwise around the seated circle, starting at lower left, are Tom Alexander, Barbara Ambler, Jewel King, Judy Alexander, Lucy Hildebrand, Bill Davis, Bill Shuford, and Walter McGuire.

George Masa—Courtesy of Barbara Ambler Thorne

A favorite trick Avery and I liked to play on guests was the panther scare. After supper, I would slip out from camp fifty or a hundred yards into the woods and by screeching while hitting my throat with the sides of my hands, I made what I considered a plausible imitation of a panther scream. Avery would tell everyone the old "painter" was back again.

Most of our guests believed it, and on one occasion, when I gave my imitation scream, a teenage boy ran for his .22 rifle and began shooting in my direction. With the bullets plinking all around me, I stopped screaming and dived behind a tree. Next morning, I scattered ketchup around the spot and took the group over to show them panther blood. To this day, that boy probably thinks that he once wounded a Smoky Mountain panther.

✣ ✣ ✣

One time I took a fisherman way up the left-hand prong to fish virgin water for speckled trout. We went about four miles upstream to where the stream and pools were almost too small for trout. When a shower started, we headed for a nearby rhododendron slick where there was an old Indian lean-to made of boughs laid over a pole frame.

When we got back to camp, I discovered I had lost the little folding leather book in which I carried fishing flies. I was sure I had dropped the book under the brush shelter, and three or four days later I went back to look for it. I carried my rifle, thinking there might be a chance for a boomer or two. Reaching a point opposite the shelter, I turned off through the underbrush. When I parted the bushes, I saw two strangers, one of whom leaned over and picked up my fly book. I stepped into view with my rifle on my shoulder and, holding out my hand, said, "That's mine."

Startled to be accosted by a rifle-carrying stranger far from anywhere, the man handed me the book without a word. A tow sack over his shoulder and another on the ground obviously contained half-gallon fruit jars, which in the mountains could mean only one thing—corn whiskey.

Neither man spoke, so I asked, "How about giving me a drink?" One reached into his sack and brought out a jar, took off the lid, and handed it to me. I took a drink; it was pretty potent stuff. The men drank too, probably to show they were not afraid of the moonshine.

Now that the ice was broken, they explained that they were bringing the liquor to the Indian reservation from Cosby, Tennessee. Each was carrying three gallons in six fruit jars. Cosby, a center for moonshine, was located in a valley on the north side of the Smokies while the reservation was in a valley on the south side. Though only about fifteen miles apart in a direct line, the two places were separated by the main range of

>>> At Three Forks, the Left Fork of Raven Fork seems relatively gentle and benign, but by the time it gets there it has descended through some of the most forbidding wilderness in the eastern United States from its source under Tricorner Knob, on the backbone of the Smokies.

George Masa—Cataloochee Ranch Collection

the Great Smokies towering to more than 6,000 feet. Cosby supplied the moonshine whiskey for the reservation and for most of the white communities within a radius of fifty miles. Generally Indians made the round trip between Cosby and the reservation carrying moonshine. This was the first trip for these whites, and they were lost. When I encountered them, they were then in the Raven Fork watershed of the Oconaluftee River when they should have been in the Straight Fork watershed. It was hardly surprising that they had gotten lost, since they were traveling through some very confusing wilderness.

The southern Appalachians can be visualized as a ladder, the Smoky Mountains lying on the northwestern leg and the Blue Ridge Mountains on the southeastern one. Between these two ranges is the great Appalachian Plateau crossed by transverse ranges running roughly parallel to each other and tying into the two legs.

At Tricorner Knob, the Balsam Mountains butt into the Great Smokies, which is the apex of the great land uplift of eastern America. Here also the two master chains of the eastern mountain system begin meshing together. Approaching each other, the two ranges change their directions often, writhing and twisting in a labyrinth of spur ridges extending out in all directions. Tricorner Knob, 6,100 feet high, is an unobtrusive hump in this jumble of peaks and ridges. Just a mile from Tricorner is Mount Guyot, a bulging hulk 6,636 feet high, only 48 feet lower than Mount Mitchell, eastern America's highest peak. Guyot is the most isolated, most difficult to reach, and therefore least known of the major Appalachian peaks.

It is surrounded by thousands of acres of forested wilderness penetrated by no roads. So dense is the spruce and fir that nowhere can one get a comprehensive view of the surrounding mountain formations. Beneath the forest is a tangled understory of rhododendron and doghobble growing over downed trees and moss-covered boulders. Cross-country travel is difficult, almost impossible. It is little wonder that for 150 years this magnificent territory has been a region of confusion, error, and even tragedy, where Indians and whites have frozen to death after becoming lost or injured.

The first recorded attempt to penetrate the area was in 1799 by a crew surveying the North Carolina-Tennessee border, which was planned to lie along the main crest of the mountain chain. Starting from the Virginia line, the survey party came down over the Stone, Unaka, and Bald mountains without too much difficulty. But when they started up into the Smokies from the Pigeon River, they quickly became entangled in "woolly heads," the laurel and rhododendron jungles so characteristic of the area. They simply abandoned the survey.

Twenty-two years later, in 1821, a second attempt to continue the line across the Smokies was successful, but this crew became confused in the Guyot-Tricorner area, losing considerable time while men descended the slopes in all directions to discover which way the waters ran and thereby determine the true crest of the range.

The next recorded penetration, in 1858, was by the remarkable Swiss-born geographer, Arnold Guyot. Like the 1799 party, he tried to come up the backbone of the Smokies from the eastern end of Pigeon River, but finding that approach too tough, gave it up.

Then he went around to the west to come up from the Oconaluftee River. With the help of three mountaineers living nearby, he made it, by following Straight Fork and then climbing to the crest of the Balsams. From there he turned north up the Balsams into the high trackless wilderness and on to the Smokies.

Guyot, the finest geographer to explore America's eastern mountains, had already measured and mapped the entire Appalachian system from New Hampshire to North Carolina. It was he who straightened out Elisha Mitchell about the correct location of Mount Mitchell, and he settled the dispute between Mitchell and Senator Clingman over which peak, Mount Mitchell or Clingman's Dome, was higher. His altitude measurements, though made with hand barometers, were remarkably accurate. Some of his mountain maps became the basis for military maps drawn up later.

But even to him, this great new wilderness was confusing, and he made a mistake, not unlike that of my two new acquaintances from Cosby. Guyot located his own mountain, Mount Guyot (so-named by another scientist), in Tennessee, on a spur on the north side of the Smokies rather than on the state line in the main Smokies chain itself, where it belonged.

The why of this error has been the source of much speculation among those who puzzle about such things. The most generally accepted theory has been that Guyot thought that the state line followed the ridge leading to Mount Sterling. But this theory can't be correct, for Guyot was fully aware of the location of Mount Sterling and the Sterling Mountain Ridge, as is proven in his notes on the geography of the North Carolina mountains; ". . . another high and massive ridge called the Cataluchee mountains also runs parallel and close to the Smoky Mts., terminating abruptly in a high sharp peak—called Mt. Sterling, not far from the Pigeon River."

What obviously happened, as Guyot's maps and notes reveal, was that Guyot mistook for the main Smoky range, a high, rugged, still unnamed spur leading northeast from Luftee Knob. Thus, he thought Luftee Knob was the junction of the Balsams and the Smokies. Thinking he had reached the main Smokies, Guyot mapped in his "spur" and retraced his steps to Tricorner to continue west. Guyot's mistake is not surprising. The high spur runs in the proper compass direction for the Smokies. Being higher, it hides the real crest beyond from anyone approaching up the Balsams from the south.

But the United States Geological Survey takes the prize for errors in this wilderness. In 1883, they began a twenty-seven-year project to map the region. The legend on the Mount Guyot quadrangle says that the area was surveyed in 1883, 1885, and 1891, with revisions in 1898, 1906, and 1909–10. Yet, even geological survey maps dated as recently as 1948 carry many glaring errors and make one wonder how much corn liquor the surveyors carried into the high country or whether they went in at all. They left out Old Black and Inadu entirely, twisting the Smokies and the state line due east from Guyot. They moved Luftee Knob, a most prominent peak, some three miles southeast of its correct location and placed it where Balsam Corner really is, at the junction of Sterling Ridge, leaving out Balsam Corner entirely. They showed the Sevier-Cocke county line

W. B. Gragg–Courtesy of the N.C. Division of Archives and History

Mitchell Falls near Mount Mitchell, highest peak in the eastern United States. Both are named for Professor Elisha Mitchell, a geology professor who fell to his death and drowned here during a scientific exploration in 1857.

as leading directly off Mount Guyot instead of off Old Black. They left out Enloe Creek, a beautiful stream some five or six miles long in the Three Forks basin. They mapped the two main prongs of Raven Fork in the Three Forks basin in such a peculiar manner that a traveler would never find where he was if he depended upon their maps.

With so many expert predecessors caught in a similar fix, it is scarcely surprising that my two new liquor-toting friends did not know where they were. I invited them to go downstream with me to Three Forks, where I would explain how to get to the reservation.

Avery was surprised to see me come into camp with the two strangers but delighted to get a drink of liquor. I tried to buy some, but the men said they had promised the whole six gallons. They spent the night, and the next morning we put them on the trail over Breakneck Ridge to Straight Fork.

Before they left, however, I made arrangements for them to bring me five gallons as soon as they could deliver their present load and return to Cosby. The price: four dollars a gallon at Cosby or five dollars delivered at Three Forks—a dollar a gallon for packing liquor twelve miles from Cosby over a four thousand-foot climb and down through that wilderness to Three Forks. Four days later the men showed up with the five gallons.

Miss Judy in the 1930s.

Portrait of a Mother

BY ALICE ALEXANDER AUMEN

A travel-writer friend once remarked to me that if anyone ever got around to writing the history of Cataloochee Ranch, most of it would have to be the story of Miss Judy. My father agreed completely; while he was generally off on his forestry pursuits, she was the one left to run the ranch—handling reservations, greeting and hostessing, supervising household and kitchen staff, shopping, meal planning, billing, bookkeeping, and promotion. In her spare time she cured hams, raised chickens, and tended a large vegetable garden.

My experiences growing up on the mountain are somewhat different from my siblings'. As the youngest, I had no recollection of Cataloochee Valley, Pensacola, Lake Junaluska, or other relatively civilized worlds. Brother Tom and sister Juju were long gone to boarding schools. Mother and Dad had their hands full with the ranch and Dad's forestry jobs, so everything for me revolved around the "extended family" of staff and guests. I never had a neighbor until I moved to New York.

We were taught early on to be self-sufficient and enterprising. I learned to be content with books, music, horses, mountains, and interesting, if fleeting, guest relationships. When Mother offered me piano lessons, I couldn't believe my good fortune.

But she did more than understand the varying interests of her diverse offspring. She was best friend as well as mother, finding ways to soothe her youngest daughter, thin-skinned and insecure, who sometimes had difficulty dealing with the inadvertent gruffness that is an Alexander family trait.

I took our education for granted, not comprehending the financial sacrifice that private schooling entailed. The cost of education was rarely mentioned, although we received regular lectures about more frivolous things such as clothes, dates, and sleeping late.

After two years as an art major at Duke University, I told my parents that I was wasting my time and their money and that I had looked into design school in New York and had even gone so far as to locate an apartment and roommates. I still find it hard to believe that there was no question about my decision to make such a major move at the tender age of nineteen. After New York and a design stint in San Francisco, I thought I was making a temporary summer move back to the mountains, but during that summer Mother and Dad made their decision to introduce skiing in the Smokies.

I never moved away again and met my future husband, another Tom, originally from Hershey, Pennsylvania, who was able to contribute other skills to the far more complex operation that skiing represented. Our son, Alex, went through an educational process similar to the rest of the family. After a sojourn of several years in Atlanta, he is back at the ranch, an admitted victim of mountain fever.

Miss Judy had made sure we all knew the basics of working in a family business: cooking, housekeeping, being polite to the guests. As I progressed into marketing, bookkeeping, community involvement, and specialty retailing, I realized the astonishing range of talents my city-bred mother had to master unaided.

Born in 1907 in Richmond, Virginia, she was one of three daughters of Dr. George Barksdale and Mary Morton Lovelace Barksdale. Whenever one of us children did anything especially stupid, mother would jokingly remind us that our grandparents were first cousins. Grandfather was very much a Renaissance man—army surgeon, obstetrician, pharmacist, farmer, inventor, artist (now collectible), and poet. Grandmother contributed the special Lovelace sweetness—and spice—to all the Barksdale offspring. I was named for the middle child, Alice. A tempestuous, titian-haired charmer and talented musician, she was, I thought, quite the most glamorous person I ever knew.

My mother, Judith Morton Barksdale, was the youngest, a lovely strawberry blond blessed with a smile that would melt Antarctica and blue eyes with a twinkle to shame the Milky Way. She was thoroughly coached in how a southern lady should run a household. In addition to household and extraordinary cooking skills, she inherited a love of gardening from her father. She would one day put all those talents to work, first in a remote fishing camp in the Smoky Mountains, then on a primitive farm, and finally on a mountaintop.

My grandfather insisted that she have violin lessons while Alice studied piano, but she really didn't like her assignment; she gave up playing her rather fine violin when we were young, only lending it to fiddler Cal Messer for ranch square dances. The two sisters often starred in musical dancing shows for World War I soldiers at Fortress Monroe, Virginia, where Grandfather was stationed. Later, at the ranch, Miss Judy played the guitar and sang mountain and cowboy ballads around the campfire. The flapper age contributed one talent I could never master; she can still perform a mean "shimmy," eyes and grin exuding pure mischief.

She was slim and athletic, being especially fond of swimming and diving competitions. In one teenage diving accident, the board flipped her against a concrete wall, breaking a number of teeth and causing a concussion and some permanent loss of hair. She learned to cover up that drawback with a becoming "wound-around" hairstyle. But finding it difficult to keep her hair in place in breezy weather, she adopted her trademark "Miss Judy hats." People my age and older have vivid remembrances of her scurrying along Waynesville's main street on her semiweekly shopping trips, always in a hat—western, fedora, beret. She protested for years that she could not wear red because of her red hair and coloring. Then she discovered the red hat. Recently one man told me that he made

Jack Szold–Cataloochee Ranch Collection

Tom and Judy Alexander in 1953 with Miss Judy sounding the rhythmic tattoo on the ancient brake drum that has always called Cataloochee Ranch guests to meals.

his first visit to Cataloochee after becoming intrigued while following a station wagon driven by a lady in a red hat. He has returned several times since.

Miss Judy had the typical army brat's geographically diverse upbringing—Richmond, Fortress Monroe, New Haven, Louisville—but spent her high-school and young adult years in Asheville where my grandfather was stationed at the Veterans Administration Hospital at Oteen. After two years at Woman's College in Greensboro, she returned to a busy and social life in Asheville. She was involved with the Junior League and worked for a while as an assistant in a photography studio. Aware that friends and acquaintances were having trouble finding attractive ready-to-wear dresses—this being before the days of half-sizes and petites—she opened a dress shop specializing in custom-made "Judy Frocks."

The story of how she and Tom Alexander met on a pack trip is told in Dad's narrative. Corbett Alexander (no kin), the pack trip's outfitter and long-time family friend, once told me that this occurred during the height of the flapper era. Evidently Miss Judy could "flap" with the best of them, but even then Mr. Tom said he could discern great depths beneath the coquettish exterior.

She and Dad were married in June 1929 in a garden wedding at the Barksdale home in Asheville. After a combination business/honeymoon trip to Nova Scotia, they returned to a rented house in Asheville. Pretty soon afterward, the Great Depression closed in.

✣ ✣ ✣

I get little input from her about this Asheville flapper-Junior-League socialite's reactions to their subsequent surroundings in the Three Forks wilderness and the isolated Cataloochee Valley. Apparently, she simply accepted it as a matter of survival in those depression times. But I marvel at her determination to take "civilization" with her into those areas. "Miss Judy," as she immediately became known in the valley, had a talent for melding old Virginia elegance with simple mountain ways and fare. We grew up with dinner candles and lovely arrangements of native wildflowers and greenery, such as galax and doghobble, on attractively set tables. She has never admitted even to dismay, much less terror, at the enormous prospect of trying to transform the derelict Fie Top farm property into both a home for her family and a retreat for urban guests.

Her physician father made sure that when she moved into the wilds, she went equipped with an elaborate medical kit and knowledge about how to use it. As word spread up the valley about her medical competence, the twenty-six-year-old girl was called on to treat everything from hangovers to broken bones to axe-cuts to "bloody flux," but she says she always drew the line at midwifery.

People looking at photograph albums of the early ranch days are often intrigued by the pictures of three spiffy black maids in uniforms, starched white aprons and caps, and the West Indian chef in his whites and high toque. When asked how these highly trained but distinctly urban serving folk were persuaded to move to these remote mountains, Miss Judy replies that room and board and a dollar a day looked pretty good to lots of people in those times.

Somehow, most everyone she has ever encountered—children, in-laws, staff, guests, local and far-flung friends—sense that anything told Miss Judy will always be held in strictest confidence. Problems, worries, and heartaches have poured into the sympathetic ears of this wise and nonjudgmental confidante, and stayed there. As one of the young wranglers said recently, "If there was more people like Miss Judy in the world, there wouldn't be no wars."

When arranging guest seating for meals, she made sure that those she regarded as "problem children" were seated next to her, while Mr. Tom would be favored with the entertaining, the prestigious and the political, and the gals on the prowl. Talk about tolerance!

Despite all that tolerance and discretion, she nevertheless has a bit of disdain for those who are supposed to "know better" or for anyone who abuses the hospitality in her home, as she still regards the ranch. While I have virtually never seen anyone angry from an encounter with Miss Judy, I have also seen this petite five-foot-one-inch lady literally turn pretentious politicos or wealthy arrivées out of her house, but without their being quite aware of what was happening. Once, at the old ranch, a famous novelist became obstreperous after partaking of a bit too much moonshine. Out he went, famous author or no, but he subsequently became fast friends of the family and even incorporated Dad and Mother into one of his books.

One evening, a prominent judge and his wife checked into one of the corner rooms of the present ranch house. They had obviously been drinking for some time before their arrival. Our outside suppers were held on the lawn not too far from their open windows, and the ranch guests were treated to a loud and profane argument. The judge finally came out to dinner, but his wife refused to emerge. Concerned, Mother went to her room and found her cowering behind the commode. Later in the evening, the judge came marching down the hall, clad only in his undershorts, demanding to know when in the hell dinner was going to be served. Of course, the next morning Miss Judy informed them that they would have to leave.

Misbehaving judge: "Do you know who I am?"

Miss Judy: "Does it really make a difference?"

I remember my sense of frustration as a teenager after a highway patrolman drove up that inaccessible road and announced that he was here to secure reservations for a "famous Hollywood personality" traveling incognito. When Miss Judy inquired the name, the patrolman said he wasn't at liberty to reveal it. To my consternation, she said that if she couldn't know the name, she didn't have any room. I never found out who the personality was, but maturity brought the realization that the actor probably wasn't really "folks," as she puts it.

On the rare occasions when new guests arrived at Cataloochee expecting a more elegant or diverting resort, Miss Judy would sense their disappointment almost immediately. With enormous tact, she'd suggest that they'd probably be happier in Gatlinburg or Asheville or elsewhere and also suggest a suitable spot, allowing guests to make their escape with both their dignity and their deposits intact.

She has had her share of wildlife experiences other than the wildcat episode Mr. Tom writes about. She always cured her hams using the "Virginia Sugar Cure," a combination of liquid smoke, salt, brown sugar, and black pepper. Each winter, she cured about two dozen, laying them out on tables in the old storehouse, which had a rear window covered only with screening.

One Sunday, our dishwasher Fred, fresh from a day off and suffering a terrible hangover, went behind the storehouse for some hair of the dog. There he confronted a large black bear standing on hind legs next to the ripped-away screen and holding one of the curing hams. Fred somehow made it back to the ranch house, stammering, "Miss Hoolie, Mi-Mi-Miss Hoolie! There's a b-b-bear out there!" We all thought it was more of Fred's hallucinating, but Mother grabbed an old pistol she kept in the bathroom and ran out to rescue her ham. She fired a shot or two at the fleeing bear, but apparently missed. Later, when asked why she kept a pistol in the bathroom, Miss Judy said that no burglar should be allowed to refuse a lady the right to use the bathroom.

In the early sheep-raising days on Fie Top, Miss Judy adopted an orphaned lamb whose mother had been killed by a bear or wild dogs. It became a ranch pet and was spoiled by all the guests, but especially by Lee Cason, the chef, and eventually the lamb itself came to be called Lee because of the way he constantly shadowed the generous cook. Of course, he one day grew to be a ram, but because of his long association with dogs, it appears that he always thought himself a dog; he would actually go out with the dogs to chase sheep, "barking" with a ridiculous baaing sound.

In those days, the ranch house had no locks; the main entrance merely had an old-fashioned wooden latch and latchstring. One cold winter night, Mother was by herself in the ranch house, with Dad away on business, and we children at school. Alone with her radio and her sewing, Mother heard rattling noises and looked up to see the latch-bar bobbing up and down. She admits to some nervousness. Pretty soon the door opened and in walked Lee (the sheep). Fed up with the snow and cold and ready for a warm fire, he had apparently figured out that pulling the leather latchstring was the way to get inside. I don't remember whether or not he was allowed to stay.

✣ ✣ ✣

Early roads and automobiles presented constant challenges. The Cove Creek road into Cataloochee Valley was winding, unpaved, and narrow, as it still is, but with no reassuring guardrails. While pregnant with me, Mother had borrowed Uncle Miller's car for her weekly shopping trip to Waynesville. On one of the harrowing curves, she encountered a Florida car whose driver had no intention of moving out of the center of the road,

Visitors into Cataloochee Valley must still negotiate the road from Jonathans Creek to Cove Creek Gap shown in this picture taken in the thirties.

so she was forced to the outer shoulder, which gave way. Her car tumbled down the mountain, rolling over about six times. She says she quickly realized that there was nothing much she could do except put both hands up against the roof of the car and let it roll. The vehicle finally came to a halt against a sturdy tree. Unharmed, she climbed out, hiked back to the ranch, borrowed Irma Bain's car and resumed her shopping trip.

The Fie Top road in the forties and fifties wasn't much better. Early guests remember the specific instructional signs posted at regular intervals: "Steep Grade. Low Gear," "If Car Chatters on Rocks, Reduce Air Pressure in Rear Tires," "Water. Cool Your Motor. Better Stop," and finally, "Last Climb. Low Gear."

In winter, the upper portion of the road, called the Dark Hollow, was usually impassable. Miss Judy was stuck in the Dark Hollow only a few times before she, Dad, and Glenn Messer developed a code. She would drive to the point where the road curves from a southern exposure to northern exposure—then and now known as the "Judy Curve." There she would pull off in a wide spot and start blowing the horn. Sound carries well in the mountains, and soon either horses or, later, a four-wheel-drive vehicle would appear to take her, children, and supplies to the top.

With their perpetual sense of responsibility to their "people" and realization that the younger generation couldn't support themselves on the summer business alone, Mr. Tom and Miss Judy pioneered skiing in North Carolina. At the age of fifty-four, she had to adapt once more to a brand-new kind of enterprise—high-volume, fast food, crazy skiers, and a whole new assortment of injuries.

For all the aggravations involved in growing up in a house full of strangers, I will always be grateful to Miss Judy and Mr. Tom for the marvelous education we offspring derived from Cataloochee's loyal clientele of politicians, artists, writers, professionals, businessmen, and simple friends and real people.

Courtesy of the Ewart M. Ball Collection, Ramsey Library, UNCA

Cataloochee Creek, in the lower gorge. Water from the creek was diverted into a concrete tunnel through the mountains to produce power at Waterville on the Pigeon River. The tunnel, constructed from 1927 to 1930, is 14 feet in diameter, 6.3 miles long and cut through solid rock—an engineering wonder for its day.

A snake fence in the meadowlands of Cataloochee Valley, on one of the old prosperous farms. Unlike most of the valley farms, which were allowed to return to forest, these fields are kept open by mowing, but now they are grazed by deer rather than cattle, sheep, and horses.

Courtesy of Great Smoky Mountains National Park

The End of the Valley

Among the first lands acquired for the proposed Great Smoky Mountains National Park were the farm- and timberlands on the watershed of Cataloochee Creek in the southeast corner of the proposed park in North Carolina. The Cataloochee Valley was occupied by nearly a hundred fine old-time mountain families. These old settlers "lived at home," keeping beehives and fruit trees. Most had a few head of cattle, many kept sheep. All farmed, raising principally corn and hay, though tobacco had become an important money crop. Their houses were well built, usually of two stories, some of logs, many of painted boards. The creek was by then a celebrated rainbow trout fishing stream, and many of the natives took in extra cash by boarding fishermen. Some citizens were known for making good corn whiskey and found ready customers among the fishermen, towns and cities in the surrounding out-country, and mountain logging camps. As mountain communities went, Cataloochee Valley was prosperous, if isolated.

The road into the valley, which the residents had built, was something of a nightmare—crooked as a blacksnake, with rocky ledges sticking up in the roadbed—stretching eleven miles across the Cataloochee Divide ridge. Bordered by cabins, tobacco patches, and hillside clearings, the road twisted its way up through Cove Creek Gap and dived into the national park, with its heavy virgin stands of chestnut, oak, poplar, white pine, and maple. The road wound down to Cataloochee Creek through dense mountain laurel and rhododendron. Just above the creek, it forked, the left fork crossing the creek and following the rich bottomland farms, the right fork climbing northward through Mount Sterling Gap into Tennessee.

In late 1932, John Needham, the acting park ranger for the North Carolina half of the park, approached me with a proposition. The Park Service had come into possession of a farm at the bottom end of Cataloochee Valley that Needham thought might be worth my leasing to convert into a tourist facility. The farm's previous mountaineer-owner, "Preacher" Will Hall, had developed the farm into a little camp for fishermen and woods lovers. With a team of mules and a drag, he had scooped out a three-acre lake in the bottom beside Cataloochee Creek. He and his wife boarded guests in the farmhouse and in two cabins he had built. If fishermen failed to catch fish in the creek, Hall charged them fifty cents apiece for trout from his lake, which he stocked from three hatchery ponds he had also constructed.

Hall's white two-story farmhouse had no running water, much less electricity or telephone. He had built a storehouse in the yard, insulated with sawdust, for canned goods, potatoes, apples, and cabbage. Directly over a fine bold spring coming out of the side of the mountain, he had constructed a rock springhouse with a series of troughs to cool butter, cottage cheese, milk, buttermilk, and cream. Above the springhouse a fieldstone wall fenced in hillside pastureland. To the right stood a big barn with log-crib stalls and a center hall, sheds on both sides for machinery, and a loft for hay and corn. Behind the farmhouse was a garden. The land had many apple trees, and along the creek were six or seven acres of good corn-raising bottomland. Woodland covered the slopes above the spread.

Needham and I arranged to meet at the Jarvis Palmer place, the first farm above Hall's, to inspect the layout. My first impression was favorable, but it was clear that many improvements and a great deal of money would be necessary to make the place suitable for outlander guests.

The first necessity was running water. About a quarter of a mile above the gate to the Hall place was a vacant house, which Needham told me I could tear down for the lumber if I would clean up the site and bury the chimney rocks. He also offered me underground piping, if I wanted to dig it up, at the abandoned sawmill village in the park at Smokemont, some fifty miles away.

Thereupon, I formed a corporation, grandly called Great Smoky Mountains Camp and Tours, and offered stock to friends willing to gamble with me on the future of a tourist enterprise to be called Cataloochee Ranch. I sold the shares for $100 apiece, requiring a $25 down payment and the balance later. Enough people signed up to make it look as though we would get the capital to proceed.

The problem was immediate cash. The deferred payments on the stock would not meet payroll nor buy beds, so I took out a personal bank loan of $250 and deposited it to the account of the corporation. Shortly thereafter, President Roosevelt closed all banks in the bank holiday, and the bank held me to the $250, refusing to allow the $250 that I had placed in the corporation account to offset the $250 personal notes.

I was acquainted with a New York man who had bought an old Asheville estate and wanted to remodel its bathrooms which dated from the Gay Nineties. He offered me the old fixtures in exchange for my help in identifying and tagging all the native and exotic trees on his estate.

I borrowed an old White truck from my friend W. F. Bridgers. Accompanied by a clever mountain boy named George Caldwell who lived on Cataloochee Creek, I went to Smokemont early one morning to retrieve the underground pipe that Needham had offered. Fortunately, we had taken along a piece of chain and soon found out it was not necessary to dig up the pipe. Knowing that water pipe came in twenty-two-foot lengths, we would measure off twenty-two feet on the ground, dig to find the pipe joint, unscrew it, hook the pipe with the chain to the truck, and pull the sections out one by one.

By late afternoon, we had several hundred feet of pipe. We tied this on one side of the truck and started back to Cataloochee over the steep, crooked road. Shortly after starting up the mountain, the old truck's fuel pump played out. We found we could keep the engine going by pouring gas into the carburetor, so I lay on the left front fender trickling gas, while George drove. We made it into Cataloochee in the middle of the night.

Judy and I had decided to turn a large old cabin on the lake bank into a dining room and kitchen; on the porch we built the ranch's single real bathroom. The morning after our late night trip, we began laying a pipeline from the springhouse to the cabin and by nightfall had running water in our new kitchen. By the end of the next day, we had plumbed our new bathroom and its fine Victorian fixtures. For lighting, we made do with kerosene Aladdin lamps, lanterns, and flashlights.

After putting partitions in the old farmhouse and cabins, we had about eight rooms, which had to be furnished. We knew of a large antique store in Asheville with much handmade mountain furniture. In its unlighted basement lay the posts and head- and footboards (but no side rails) for dozens of old rope-spring beds, which were priced at a dollar a set. With flashlight and pocket knife, I went among these beds, scraping down through several coats of varnish or paint to pick out desirable species of wood. I selected eighteen walnut, maple, cherry, and poplar beds built in almost as many different patterns as there were beds.

Mr. Bridgers let me use a corner of his large garage as a workshop for removing the varnish and paint, a tedious job involving chemical remover, steel brushes, sandpaper, and emery cloth. After several weeks of work, I had the headboards and footboards fairly clean and then got a cabinet shop to make side rails and alter their sizes—doubles, three-quarters, and twins—so they could be fitted with modern springs. Some of these old beds were still in use at the ranch more than half a century later.

At the same antique shop, we also picked up tables, dressers, chairs, lights, firedogs, fire shovels, pokers, and so forth. We equipped ourselves as a mountain farm by buying a big old work horse and harness plus twelve hens.

Though we had no money for riding horses and saddles, I went to Lush (pronounced "Loosh") Bramlett, who ran the livery stable in Waynesville, and outlined our proposal for a ranch in the park. I explained that we needed horses.

Lush looked at me a minute and asked, "How many horses do you want?"

I replied that six would do for a start.

"When do you want them?" he asked.

"About May 1," I said, it then being March.

"Well," he said, "I couldn't truck horses all the way across the mountain to Cataloochee, but if you will bring two or three boys over the mountain to the Cove Creek post office about eleven o'clock on May 1, I'll meet you there with six horses, saddles, and all the gear. You take them, feed them, shoe them, and keep them all summer. Keep a record of the money you take in and at the end of the summer, give me half. If a horse dies, drag him off, and bury him, and that'll be the end of it."

Lush and I operated under that verbal agreement until he died in 1940. There was never a misunderstanding or harsh word between us. I learned over the years that he dealt that way with everyone, mountain folk, city folk, or politicians. Though he neither drank nor smoked, he liked good food and had an abdomen shaped like a full moon. His only other dissipation was chicken fighting, on which he bet heavily. His country friends raised fighting cocks for him on their farms.

To house Lush's horses, we installed partitions in the machinery shed running around the barn, making stalls for more than twenty horses.

✣ ✣ ✣

We soon found out how right Judy's father, Dr. Barksdale, had been when he told me we would find it far less expensive to live in the country than in the city. By that time we were feeding another mouth, three-year-old Tommy, but there were no telephone, light, or dairy bills. We enjoyed fresh vegetables from the garden and apples, berries, and grapes from our trees and bushes. We caught delicious fish in the stream and lake. We canned everything possible, including meats, sausage, fruits, and berries, for wintertime.

One of my vivid memories is Judy with her homemade canning device, consisting of two great galvanized washtubs, sitting over an outdoor fireplace like those the moonshiners made for their stills. Fruit jars full of produce were placed in eight or ten inches of water in one tub, covered with the other tub turned upside down, and steamed for the required time.

Our mountain neighbors in the valley were delightful, and one, Jim Caldwell, who lived about five miles up the creek, became our ranch foreman. Jim was a tall, fine-looking man—a favorite of the guests, particularly the children. Before we came to Cataloochee Valley, his principal cash income, as with many of his neighbors, derived from corn whiskey and farming. His wife, like the other women, spun her own wool and wove her own cloth. Their son, Larry, eventually succeeded Jim as foreman.

In the first occupied house up the creek above the ranch lived Harley and Jarvis Palmer. Harley was a heavy-set man, a heavy drinker, and good friend of ours. His brother, Jarvis, one of the most substantial citizens in the valley, had a great many cattle, which ranged every summer on the high mountains at the head of the valley. Riding into our ranch one day on his white mule, Harley told Judy that he had been on a week's drunk and wanted to know if she had any whiskey he could sober up on.

Judy had just made wine from grapes growing near the house, so she filled a pint fruit jar and handed it to Harley, who started home on his mule with the jar in his coat pocket. In about fifteen minutes, he came riding back, with the top and broken neck of the fruit jar in hand, saying he had dropped the jar on the road and wondered if Miss Judy could give him a little more. She did, of course.

We later found out that Harley had drunk the wine as soon as he was out of sight and then broke the jar, judging that one pint of grape wine hardly sufficed to sober a man up from a week's drunk on real mountain whiskey.

A Richmond-born doctor's daughter and Junior League belle, Judy Alexander had to acquire a lot of new skills in a hurry when she moved to the Cataloochee Valley. Here she demonstrates her washtub canning technique. Her son Tommy, then about four, remembers that she usually kept a switch tucked in one of those high boots as a disciplinary reminder.

Aunt Maria Palmer, Jarvis and Harley's spinster sister, lived with them and made beautiful patchwork quilts in traditional mountain patterns: "Double Wedding Ring," "Bird in the Window," "Flower Garden," "Texas Star," "Trip Around the World," and "Nine of Diamonds," many of which we bought and still own.

Aunt Maria always wore a great bonnet to protect her face from the sun and urged Judy to wear one. Judy learned the habit to her advantage, since with her fair complexion she sunburned and freckled easily.

The most engaging of the numerous Palmer family was "Turkey George," who lived far up the valley. Throughout his lifetime, Turkey George was the valley's most dedicated hunter. His family practically lived on the meat from the 106 bears he is said to have killed during his lifetime. He died in 1944 at the age of 87. His old homeplace, near the mouth of Pretty Hollow Creek, is a favorite camp site for ranch pack trips.

It has always been traditional for mountain people to help each other out. Every citizen of the valley assumed that he could count upon help from fellow citizens in communal "workin's" such as barn raisings or roofing projects. Hog killings were also festive occasions, as were corn shuckings, where a jar of whiskey, hidden under the enormous mound of ears of corn, served as an incentive for faster shucking and a guarantor of good times afterward.

When the valley residents discovered that Judy had learned some of the essentials of medicine from her physician father–stitching wounds, taping broken ribs, and giving paregoric for intestinal disorders, and so forth–she was frequently called upon. In turn, she learned some mountain remedies, such as tobacco poultices for bee stings.

Our lease from the National Park Service–a year-by-year arrangement–discouraged us from spending much money on major improvements, but we soon found the ranch too small and crude to pay off. The Park Service gave us permission to dismantle three or four one-room hewn log cabins in the valley, which we reassembled as additional accommodations, adding a pioneer atmosphere.

High up on the pasture hill above the farm perched a tiny cabin we called the Crow's Nest, very popular with guests despite the steep climb. One guest, Mrs. Don Fitton of Hamilton, Ohio, left her box of face powder open on the dresser one day, and a milk cow wandered in to powder her nose and deposit a cow-pie on the floor for good measure. I remember Judy climbing those hillside steps with a bucket of hot water and mop to clean up the mess.

Families in the valley helped us build several wood-floored tents along the banks of the creek as additional quarters, providing accommodations for about twenty-five guests in all. We eventually installed a bathroom in the upper farmhouse, bringing in water from Nelson Branch behind the house. Though temporary, it worked, giving us a grand total of two bathrooms—plus the creek, of course, which many of us preferred for afternoon bathing.

✣ ✣ ✣

In the spring of 1933, Cataloochee Ranch opened its doors to guests, the only tourist facility on the North Carolina side of the new park, then receiving widespread publicity as the first national park in the East. We put copies of our new brochure in hotels, park headquarters, ranger offices, and other places where potential guests might be. Avery Gouge, our old woods cook, was the first chef. We later located a chef and three maids in Asheville who did a fine job. A native of the West Indies, Archie the chef not only was an excellent cook but knew something about horses.

After working hard throughout the late winter and spring to get the ranch ready, we were resting on the front porch of the old farmhouse one Sunday afternoon in April. A big, shiny, black automobile chugged into view, inching along the wagon track covered with creek gravel. Walking in front of the car and bending to throw the largest rocks out of the road, was a fashionably dressed woman who turned out to be Mrs. James Barker of Chicago. She and her husband, the vice president of Sears, Roebuck & Co. in charge of retail stores, were our first guests. That delightful couple got us off to a fine start in the resort business and eventually became close friends, returning to Cataloochee Ranch many times over the years.

At the time of their first visit, the Sears, Roebuck catalog ranked second only to the Bible in literary esteem in the Cataloochee Valley, and word quickly spread that "the Roebuck man" was in residence at the ranch; kids of all ages hid behind bushes and trees to get a glimpse of him.

We had started the ranch enterprise in the bottom of the Great Depression at a time when the New Deal was stimulating a new interest in conservation activities of various sorts. Foresters were suddenly in great demand, and I began receiving job offers from federal and state agencies.

Offered the job of assistant forester of the Tennessee Valley Authority, I went to Knoxville to interview the TVA's chief forester and the TVA board. I became acquainted with David Lilienthal, then a board member, subsequently one of the early guests at the ranch,

Archie, the West Indian chef, ringing the brake-drum dinner bell at the old ranch in about 1934. In those hard times, the ranch had no trouble getting highly trained professional kitchen staff to take up residence in the wilds of the Smokies. With the coming of the war and relative prosperity, it became much more difficult.

and later chairman of the U.S. Atomic Energy Commission, among other things.

I got offers from four state forestry departments, but finally accepted a post in the Civilian Conservation Corps program as assistant district inspector for the southeastern district, encompassing six states. The CCC had been created to employ out-of-work young men on various outdoor projects, such as building roads and trails. The chief inspector was C. F. Evans, an old-time U.S. Forest Service man I had known for years. We were to supervise the installation and operation of 106 CCC camps working in our district.

The new job brought in income to help put the newly established ranch on its feet, but it kept me from operating it. My brother Miller, then in a law firm in Houston, loved the mountains as much as I and dropped everything to come help Judy. I commuted nearly fifty miles each way from Cataloochee to my office in Asheville, driving home over the crooked road at night, often in heavy fog.

Tom Alexander's brother Miller abandoned his Texas law practice to help Miss Judy manage the old ranch. Here he holds several saddled horses during a lunch-break in the lush pastures of the Purchase, on the Cataloochee Divide. The present owners of the beautiful Purchase property, located close to the present ranch, are donating it to the adjoining Great Smoky Mountains National Park.

Tom Alexander's vision for the Cataloochee Ranch was a jumping-off place for city folks to get a taste of the mountains that had so infected him. Horseback and pack trips appeared the logical way to travel. At the time, the Civilian Conservation Corps was building a network of trails throughout the new national park, most of them elaborately engineered to be suitable for horses. The American Forestry Association, which up till then had sponsored pack trips in the western United States, approached him in 1935 with a proposal to conduct ten-day-long trips throughout the park. The proposal had the enthusiastic support and cooperation of park authorities. The above picture shows members of the first AFA party at the old ranch. From left to right are AFA representative C. H. Collingwood; Corbett Alexander, an experienced pack-trip outfitter and longtime friend of Tom's; Lillian Judd; Tom himself; Edith Foster; Gladys Hendrie; Marian Mair; Lenore Suder; a Mr. and Mrs. Zimmerman; T. B. Saunders; Jim Lite of the National Park Service; and Avery Gouge, cook.

Many trout fishermen came to fish Cataloochee Creek, and if they had no luck in the stream, we'd let them catch a few from the lake, charging them fifty cents apiece so they wouldn't have to go home empty handed. Besides fishing, the principal entertainment at the ranch was riding and overnight pack tripping. On our first overnight rides from the Cataloochee Valley, our gear consisted of ponchos, blanket rolls (we had never heard of sleeping bags), and tent flies for shelter.

After our first season of operation, the American Forestry Association asked us if we could handle as many as twenty riders on a ten-day pack trip through the Smokies the following summer. We agreed to undertake this ambitious Trail Riders of the Wilderness trip, the first of many we were to guide over the years. With packers and cooks, some of the parties would consist of as many as thirty people, requiring forty-five to fifty horses and mules to handle riders, duffel, and food. I knew the superintendent of a CCC camp on Big Creek who had had a great deal of pack-train experience in the West and who volunteered to help us load up the pack horses.

On one of the first trips, we camped on the Cherokee reservation, and several Indians came into camp to visit. One, named Will West Long, well known for the Indian dances he put on in Asheville and other towns, volunteered to organize a dance for us that night.

Tom would never leave for the woods without his ancient "pack-trip hat," veteran of many a storm.

Our riders joined in the Friendship dance and the Green Corn dance.

Among the riders was a dark-skinned girl named Grace Price, who looked quite a bit like an Indian herself. During the evening, she told Long that she had always wanted to learn the Cherokee language, and the next morning well before breakfast, Long made an unplanned visit to camp to begin her lessons. He walked over to Grace rolled in her blanket, and with heavy fog hanging over the campground, told her it was time to get up. Waking to the sight of an fog-shrouded Indian bending over her, Grace almost jumped out of her skin.

That day we headed for my old Three Forks camp where we planned to spend our second and third nights. We took a very rough trail from the Big Cove campground in the reservation up Toe String Creek and along Hughes Ridge to the old trail that we traveled when I had my camp there. To supplement my own string of horses, I hired two Indian boys with mules to help us pack horse feed into Three Forks. The boys went ahead, missed the trail and followed Hughes Ridge all the way up to the Appalachian Trail on the main crest of the Smokies. Realizing their mistake, they backtracked to find our horses' tracks where we had turned off on the Three Forks trail. They finally reached camp in time for the horses to have some oats before dark.

At Three Forks the next morning, June 22, 1934, we had a heavy frost, and our group thought they would freeze to death. But after the sun came up, a good many mustered enough courage to swim in the ice-cold water of the Three Forks pool.

Pack tripping proved very popular—and still is. Whenever guests want to go out for overnight or longer trips, we organize trips to regular campgrounds, some located on peaks for good views, some down along the streams for trout fishing. Sometimes we camp at Park Service shelter cabins where black bears frequently come into camp to steal food.

While bears are usually not dangerous to people, they do considerable damage to food and camping gear. On one pack trip, a woman guest had placed her sleeping bag a little way from the main group, her riding boots alongside, with a plastic bottle of perfume tucked inside one of them. That night, evidently wanting salt, a bear stole into camp and bit right through the leather boot, perfume bottle and all, getting scent all over its face. We ran the bear off, but whenever he came near camp, we knew he was there by the smell.

Bears can be terribly artful in stealing food. We used to pack food in canvas bags buckled shut and hung on pack saddles. In the beginning, when we reached camp, we

Above, wrangler Larry Caldwell leads a pack string past Uncle Steve Woody's house in the early thirties. Below, riders pause for lunch atop Charlies Bunion on the North Carolina-Tennessee state line. According to one theory, Horace Kephart named the peak for a sore-footed companion during a 1929 mountain-roaming expedition in the Smokies.

Uncle Steve Woody, a prominent Cataloochee Valley resident, whose house in the background is one of the few left standing. Though it was covered with wooden siding as the family prospered, the Woody house began life as a log cabin, as did so many of the prosperous homes in the mountains.

Courtesy of Great Smoky Mountains National Park

often put these bags on the ground without unpacking them but discovered that bears would simply carry the entire bag into the nearby woods to get at the bacon, butter, or bread inside. One night, we covered two bags with a tent fly guarded by two men sleeping on either side, but a bear managed to tear open the heavy canvas and steal the bags without waking the guards. Bears will even carry off sacks of oats, apparently out of sheer cussedness, since oats are not ordinarily part of their diet.

During a luncheon break at Ice Water Springs, one rider, a doctor, went into the woods to answer nature's call. Feeling pieces of bark falling on him, he looked up and saw a bear staring down from a tree limb. We saw the doctor hopping into camp, pants still dragging around his ankles.

One day, David Lilienthal, Dorothea De Swienitz, Elizabeth Roby, and I headed for Flat Creek, a fine campground in the Balsam range at the head of Cataloochee Valley. We rode up the valley for five miles to the homestead of Steve Woody, a prominent Cataloochee citizen with a quick sense of fun, and then in his eighties. Finding Uncle Steve sitting in a homemade chair in the yard, we tied our horses to his picket fence to rest and water. Practicing common mountain courtesy, I asked Uncle Steve to come with us. Without hesitating, the old man replied, "Wait till I get my coat and saddle my horse." And off he went with us for the three-day trip.

Another time, I rode into Uncle Steve's yard with some riders after a downpour and asked the old gentleman if I could build a fire to dry out. Uncle Steve said I was welcome, so I climbed the rail fence into the old stack yard where plenty of old chestnut rails were lying about. I landed right in the middle of a large nest of yellow jackets that quickly covered me up. Slapping and flailing, I made it back across the fence. When I could catch my breath, I asked Uncle Steve why he hadn't told me about the yellow jackets. "I thought you'd find 'em," he said.

It wasn't until he was on his deathbed years later that Uncle Steve told the whole story about the Beech Grove School. Around 1907, wanting a new school to replace the tiny log structure they had, the Cataloochee Valley residents sent a delegation of three, including Uncle Steve, to talk to the Haywood County Commissioners. The men were turned down flat, and on their way home across the long, winding road, stopped to drown their sorrow in a jug of liquor. Knowing they wouldn't receive a welcome return without the promise of a new school, they formed a plan and made a vow. That night, the old school burned to the ground. The new Beech Grove School was built in 1908. The vow the conspirators made was that the last man alive would tell who set the fire. That man was Uncle Steve, who finally revealed that he was the culprit.

Before our arrival in Cataloochee, Uncle Steve's son, Jonathan Woody, had left the valley to work for a bank in Atlanta. (Jonathan later became president of the Bank of Waynesville, a prominent figure in national banking circles, and one of my most cherished friends.) During our first summer there, Jonathan drove home from Atlanta for a vacation in the valley, where he picked up all the gossip about the new ranch down at the lower end of the valley and those Alexanders who ran it. My first encounter with him came one morning as I was leading a party of riders up the road and met a car—a great rarity in the valley—coming down the narrow thoroughfare. We stopped, the car stopped, a city-dressed young man got out, and with no further prelude said, "I'll trade you a drink of red for a drink of white."

I said nothing but reached into my saddle bag, pulled out my little Mason jar of corn, and we solemnly exchanged drinks there in the middle of the road. Thus began our deep friendship.

Among our favorite camping destinations was Mount Sterling, where there was a log cabin with a water supply and a steel lookout tower from which we could see a broad

A favorite camping spot for ranch pack trips has always been Mount Sterling, at the northeastern end of the Great Smoky range. Once open pastureland, Mount Sterling has reverted to forest since the national park took over, as have most of the Appalachian "balds" once thought to be natural phenomena. Some experts now believe that ancient Indians first cleared these balds many thousands of years ago and that they have been kept open ever since through grazing.

The house of "Boogerman" Palmer, near the Caldwell Fork of Cataloochee Creek. Palmer, whose given name was Robert, acquired his nickname as a shy schoolboy; when asked his name, he would hide his face and mutter "boogerman." Though the house vanished long ago, Palmer's former holdings contain some of the most beautiful virgin timber in the national park. During the logging boom that stripped most of the Smokies, he refused to let loggers touch it.

panorama of the eastern end of the Great Smoky Mountains. Another was Ledge Bald, a grassy knoll on the Balsams, where the mountaineers ranged their livestock. Purchase Knob, a grassy pastureland on the southeastern border of the national park, was part of the divide between Cataloochee and the agricultural lands of Jonathans Creek.

Coming down the trail from an all-day ride to the Purchase late one afternoon, an empty burlap sack that had carried corn to feed the horses slipped off my saddle. As I caught the sack on my toe and tried to kick it into my hand, the horse shied and threw me into the deep weeds on the bank, momentarily stunning me. A guest named Marcia Lamb, who is still a regular guest at the ranch, turned to find out what had happened. Seeing my empty saddle, she experienced what she said was the greatest fright she had ever known.

We often camped at Beech Gap near Ledge Bald, close to an old logging railroad that came up from Round Bottom to the crest of the Balsam range. Living there in a small shack was Charles Ray Caldwell, who acted as ranger for the lumber company that worked the fine Cataloochee timber. One day, a party of riders and I arrived at his camp just as he finished cooking a groundhog, which he insisted on sharing with us. The animal was delicious, though many of the city slickers thought it bore too strong a resemblance to rat.

William A. V. Cecil—Cataloochee Ranch Collection

From the opening of the first Cataloochee Ranch until his death in 1978, fiddler Cal Messer was a dominant and greatly beloved performer at the weekly square dances. Two mountain ranges removed from the ranch, Cal would leave afoot on Thursday to arrive in time for the Saturday night dance, then depart Sunday morning to get back home by Tuesday to tend his tobacco. A few years after the new ranch was established, Cal and his family moved to Maggie, greatly shortening his weekly commute.

We introduced our guests to square dances in our dining room, the musicians all being neighbors, except for the fiddler, Cal Messer, who walked or hitchhiked from Cosby, Tennessee, thirty-five miles away. Cal was slowed by a club foot. On the Thursday mornings preceding our Saturday-night square dances he would leave Cosby in time to make it to the dance. He would start walking back to Cosby after the dance, arriving there the following Tuesday. If we had a square dance scheduled the following Saturday, he would start out again on Thursday. A notable fiddler, Cal composed many comic songs and fiddle tunes that still get played when string bands assemble, and once or twice he was invited to perform up north.

Usually plenty of moonshine whiskey was on hand at the dances. One Saturday, Jarvis Caldwell and his wife went off to visit relatives, forbidding their three boys to go to the dance. But one of the boys, named Herschel, rode one of his father's horses to the dance anyway. When Jarvis got home next day, he spotted that horse's distinctive prints in the dirt and followed the tracks to the ranch. Herschel got a whipping.

At the end of that first Cataloochee summer, we rented a house in Asheville and moved into town for the winter, while Miller and one or two mountain boys stayed at the ranch. Later on, we spent several winters in the valley and found the cold but sociable winter months to be some of the most enjoyable times of all.

✣ ✣ ✣

During the five years we lived there, we saw many fine mountain families, our good friends, forced to move away as the federal government consolidated its holdings in the new national park. It was heart-rending to see these excellent people, some crying, with their household goods piled in wagons, leaving their beautiful valley.

The house of Lush Caldwell on Messer Fork in the Cataloochee Valley. Like most of the 139 Cataloochee homesteads, this attractive structure was burned by order of the Park Service—an outrage in the eyes of many former residents and modern visitors alike. Cades Cove, a similar isolated mountain community where quite a few of the old houses were left intact and on display, is now one of the most popular tourist attractions in the Great Smoky Mountains National Park.

Courtesy of Great Smoky Mountains National Park

This miniature "trail of tears" was comparable to another disgraceful episode that had occurred a century earlier, when the Cherokee Indians were forcefully relocated from the southern Appalachians to Oklahoma. Only those clever enough to hide from the U.S. soldiers remained to form the nucleus of the present Eastern Band of Cherokees. To this day, the Eastern Cherokees still seem more resourceful than their transplanted brethren. In contrast to western Indian communities, many of which are mired in poverty and alcoholism, the North Carolina Indian nation is a thriving entrepreneurial community, prospering largely on tourism.

The nineteenth-century Cherokees were driven from their ancestral mountain homelands because outlanders wanted to farm their lands; the white mountaineers were forced out because outlanders wanted to hike and fish in their mountains. Just as government representatives deceived the Cherokees, state agents negotiating in the name of the national park deceived the whites. They promised the Cataloochee people that they could live in the valley for the rest of their lives. But no sooner had the mountain people deeded their lands than officials began instituting restrictions that forbade hunting, grazing of livestock, cutting green timber, or planting gardens on sloping land. Fishing was only permitted under rigid restrictions. The Park Service, in short, made it impossible for these self-sufficient people to earn a living in the only ways they knew, despite their "lifetime" tenancy.

Other government agencies that bought or condemned land in the southern Appalachians had much more humane policies. The U.S. Forest Service never condemned lands for national forests but cleared titles on lands donated voluntarily. The TVA was the most humane of all. In procuring farmlands to be flooded by lakes, they set up a real estate department to help farmers buy other farms and even provided transportation so they could inspect the lands and negotiate for them. Supposedly the most conservation-conscious of agencies, the National Park Service has the blackest human conservation record of any government agency operating in the Appalachians.

Not long after taking over Cataloochee, for instance, they systematically and arrogantly began burning or dismantling the human dwellings—log cabins, frame houses, barns, smokehouses—that amounted to rare relics of nineteenth-century frontier life virtually vanished from America. Though a few buildings remain at Cataloochee—and though the Park Service is now belatedly trying to conserve these human artifacts, as valuable and interesting in their own way as the park's natural glories—the genuine character of an utterly self-sufficient human community is gone.

Over five generally happy years at Cataloochee, we too grew increasingly hampered by Park Service restrictions and the year-by-year lease arrangement that discouraged us from improving and enlarging Cataloochee Ranch. So I accepted an offer in Pensacola, Florida, to acquire and manage pine timberlands for Philip G. Rust, a wealthy acquaintance whose wife Eleanor was a daughter of Irené DuPont. The Rusts were to become close friends. We held an auction to dispose of livestock, farm equipment, and household goods, including wash bowls, pitchers, and lamps. Enameled chamber pots sold briskly,

Courtesy of Great Smoky Mountains National Park

A view of the old Cataloochee Ranch in 1940, after the Alexanders had departed but before the buildings had been demolished. Already evident in this picture are the young pine plantings that eventually transformed these clearings into deep forest.

fancy innovations to people who had always depended on trips to the outhouse. Having burned our bridges behind us, we packed our furnishings and, with our six-year-old son Tommy, three-year-old daughter JuJu, infant daughter Alice, and dog Foxie, moved to Pensacola.

Siddell Studios–Cataloochee Ranch Collection

From Walker Bald (before it was overgrown with brush and timber), the view of Maggie Valley (before it was overgrown with people and houses). Tom and Judy Alexander in the foreground.

On the Divide

"How did you ever find this place?"

Hardly a person negotiates the steep, winding road up to the present Cataloochee Ranch without asking that question, for the ranch is well hidden; from the valley, there are no indications of habitable land on these ridges.

By the second spring on the Gulf Coast, I could stand separation from the mountains no longer. A year in the west Florida flat country had reaffirmed my conviction that the southern Appalachians are the finest place in the world to live.

My business was promising, our home overlooking Pensacola Bay was pleasant and comfortable, our neighbors were charming. Those pleasurable and easy recreations of the Gulf Coast country were ours also—bass fishing in the clear freshwater streams, excellent sea trout fishing in the brackish bayous, game fishing in the gulf. Dove and quail were plentiful on the thirty-five thousand acres of forest property I managed, and there were even turkey and deer. In town, my absent friend's fine saddle horses in need of exercise were ours to ride; his sailboat was at our disposal. Golf on the breeze-cooled courses was cheap. Friends were always ready to play, and competent caddies were eager to improve a lie or trim a score. For anyone who has not contracted mountain fever, it may be hard to understand why someone in the prime of life would toss all this overboard and move to a log cabin on top of an inaccessible, windswept mountaintop.

While living at the old Cataloochee Ranch, I had heard vague stories about Fie Top Mountain, just over the Cataloochee Divide, consisting of a deadening, a cabin or two, and a farmer who raised bountiful crops of potatoes and kept cattle and sheep. I had often tried to find time to go up there and take a look but never had.

Still, my imagination could supply some of the missing details: a deadening—an area of trees killed by girdling—meant new ground, probably trashy, but surely rich and not eroded; cattle and sheep meant grass; potatoes meant a high, cool climate; inaccessibility probably meant uncut timber; cabins meant unspoiled and independent mountain people. I carried this vague picture in the back of my mind for several years.

She has never said so, but I have little doubt that Judy saw the symptoms of my mountain fever relapse the winter before. The old ranch album would come out and every picture in it—of old neighbors, big timber, log cabins, and the stream—would be lingered over and talked about as exhaustively as fine wines by connoisseurs. Foxie, the dog who had been with us at the old place, would be taken up and fondled. Hunting and fishing episodes would be recounted. Horseback mishaps would be revived and enlarged upon. Even the drunks that we had to corral and drive off after the square dances were lovingly remembered, while hardships such as storms and automobile accidents, financial problems, and droughts were recalled with nostalgia. Such are the symptoms of mountain fever.

As things turned out, though, it wasn't mountain fever but jungle fever that prompted our return to the Smokies. My little forestry practice in Pensacola was going well; I was managing or consulting on large tracts of timber for several clients and traveling throughout the Southeast. One evening in early 1938, while spending the weekend in Alabama at the plantation home of one major timber owner, a telephone call came for me from Wilmington, Delaware. The connection was bad, but I gathered that a man was wondering when I could get to Panama City to look into some logging prospects there.

I told him I was on my way to North Carolina at the time but would be back in Pensacola Wednesday night and could therefore drive to Panama City Thursday morning. Through the static I then understood the man to say that I would have to wait, then, until the following Saturday but that he was disappointed, since he had already made reservations for me on the Tuesday plane from Miami. "Our president is there now, and would like to be there when you make your inspection," he said.

"But I will drive over," I protested. "It's only a hundred miles from Pensacola. I won't be in Miami."

"The planes only go on Tuesdays and Saturdays," he answered.

We continued arguing in this confused vein for several minutes more until it finally dawned on me that he was talking about Panama City in the Panama Canal Zone, not Panama City, Florida. I told him I would change my plans and be aboard the Pan-American Clipper flight from Miami the following Tuesday.

After a series of buses, automobiles, and planes and hasty efforts to get shots, travel papers, and money, I managed to reach Miami in time for the Tuesday departure of the Pan-Am Clipper. I was always captivated by the romance of those great seaplanes, the pioneers of transoceanic air travel. When we arrived at the air terminal, the clipper was moored in the water next to the shining, all-metal terminal building, its four motors idling, workmen storing mail and packages aboard. Reservations and papers checked, we passengers climbed down the gangplank through the hatch in the top of the fuselage, settling into the facing seats in the four Pullman-like compartments. We took off in stages, lumbering through the water at first, then planing on the "step," and finally taking to the air in a plume of spray.

After stops at Havana and Kingston, and a change to a land plane in Barranquila, I was met at the Balboa airport by a Mr. Sexton in his chauffeured Packard. Sexton was the president of the company's Latin American timber operations, and as we motored into town, he filled in the details of my mission. His company had a concession from the Panamanian government to harvest timber on large government holdings along the isthmus. In the course of their operations, which had consisted mainly of cutting Santa Maria, a tropical hardwood not widely known in the United States, the company had unexpectedly found mahogany in remote regions of the forest. They were trying to raise capital for the machinery needed to log it. I was to survey their operations and advise potential investors as to whether the expenditure was justified.

We started our survey by sailing down the coast for nearly two days in the company's motor yacht, arriving late one afternoon at a sheltered bay and a small clearing in the jungle with a sawmill and fifteen or twenty palm-thatched huts. After landing and inspecting the mill and village, we spent the night aboard the boat, then proceeded further down the coast for two planned penetrations, or jungle trips. The first of these went off uneventfully; we found a little beach and landed with several native guides and enough guns, knives, water, cigarettes, etc., to make me feel like Livingston in Africa.

The jungle-covered mountains rose sheer off the beach and we climbed for several hours straight up. The trees were large and of many species, with an overlapping canopy, and the underbrush, primarily of large palm, was thick. The natives carried machetes to trim a way. On the whole, however, it was no more difficult to negotiate than our Smokies. I was fascinated by the abundant wildlife and gorgeous birds, chiefly parrots and macaws, and the apparent absence of snakes.

Sailing further down the next day, our party of three whites and five natives began our second penetration. By afternoon, after hours of thrashing across the jungled mountainsides, we were unable to locate the expected trail leading back to the spot on the coast where we were to rendezvous with our boat. We dispatched a couple of natives to find the way back and to tell the captain to send in water and food. Later, as we concluded we were really lost, we sent two more guides on ahead to look for someone with a boat who might take them to our ship and have it brought around to where we would probably come out.

That left us with one native, a fine old black Venezuelan who became my pal. As we made our way down a rocky, dry riverbed, I shot four wild turkeys with the shotgun I was carrying. Night caught us at a falls, where we stripped and bathed, while the old guide split two of the turkeys and spitted and roasted them. Even without salt, and tough as leather, they tasted delicious. We slept on a gravel bar, fighting mosquitoes.

We broke camp about daybreak without breakfast and continued down the river all morning. I shot an agouti—a large rodent-like animal—and saw a howler monkey and tracks of pumas, tapirs, deer, and wild pigs. We finally struck a native trail at about 1:00 p.m., and an hour or so later met our ship's captain coming up the trail with food and water. Our fun was over.

Despite all the quinine I had taken, I came down with malaria shortly after returning to Pensacola. I was so sick that doctors told Judy that I might not survive. They asked and got permission to extract some of my blood to inject into insane people on the theory that the shock of the malarial infection would jolt them back to sanity—if it didn't kill them.

One day I awakened from a feverish coma obsessed with images of the Smoky Mountains. I decided that as soon as I was able to travel, I would head for the mountains and look around.

In mid-May, I drove to Asheville and for three days rambled through the hills and hollows up the side roads that led into coves adjoining the Great Smoky Mountains National Park and the national forests. A relapse of malaria cut these expeditions short and threw me into the Asheville hospital for a couple of days, after which I returned to Florida. These relapses occurred frequently from then on.

Then came July, when every resident of the inland Deep South who can afford to vacates to the mountains or the beach. I packed Judy and the kids in the car, and we drove to Asheville, arriving the second day in time for the family to take the afternoon train for Richmond. There they would visit Judy's mother and father for a month, while I reveled in the arms of my mistress, the Carolina mountains.

On my May trip, I had scouted several places that might make an attractive resort, most of them located in the valleys along streams offering fishing or swimming. A prime requirement was access to the national park or one of the national forests so that people could enjoy overnight horseback trips in unbroken expanses of forestland to experience the sense of true wilderness. Even more than water and forests, however, I wanted altitude. Though beauty is a magnet and recreational opportunities desirable, the main reason people vacation in the mountains is to escape the heat of the cities.

A rule of thumb is that temperatures, day or night, will average about five degrees cooler for every 1,000 feet of altitude. At 2,000 feet, you may be sweltering in ninety-five-degree heat, but climb to 5,000 feet and enjoy maximum temperatures of eighty degrees. What a difference that fifteen degrees makes!

Another aspect of high elevation is that while the precipitation will usually be greater than lower down, there is actually less fog; morning after morning, valleys will be overcast and gloomy, while the peaks stick up through white oceans of clouds like islands, basking in vivid sunlight.

Another blessing of the southern Appalachians is the quantity and quality of water. Located about equidistant from the Gulf of Mexico and the Atlantic, the southeastern mountains receive abundant year-round precipitation. Great revolving weather systems sweep up enormous quantities of water vapor from the warm gulf and the Atlantic and carry it up the slopes until it cools and precipitates over the tops.

Most of the moisture percolates into the ground, where it collects in underground channels that eventually join and find outlets lower down in the form of springs. The peaks and ridges act as sponges, holding great quantities of purified water and gradually releasing it so that the springs continue to flow long after precipitation has ended.

Abundant precipitation and relatively benign climate accounts for the Smoky Mountains' heavy and enormously variegated growth of timber, flowers, and shrubs. The southern Appalachians contain more species of trees than the entire continent of Europe,

A common wintertime occurrence in the Smokies is the temperature inversion, in which colder air settles into the valleys, producing a sea of clouds and leaving the higher elevations clear and basking in relative warmth.

more than the entire northern half of the United States from the Atlantic to the Pacific. The luxuriant floral growth includes many species of rhododendron, mountain laurel, azalea, and hundreds of flowers on forest floors, open meadows, and mountain balds. Many indigenous trees also produce magnificent flowers—tulip poplar, three kinds of magnolia, silverbell, serviceberry, dogwood, hawthorn, locust, and redbud, to name a few.

The morning after the family left for Richmond, I drove straight for Fie Top. I knew where the road turned off from the Soco Gap highway. Little more than a sled trail, it was fortunately dry, or I would never have made it over that first red hill, with its 20 percent grade. The trace wandered over rises and skirted a branch tumbling over the rocks, winding past dilapidated little log cabins. I passed a man or two, and through my rear view mirror could see them turn to stare at my Florida license plate, wondering what an outlander was doing on that treacherous mountainside.

Halfway up, at the top of the particularly steep pitch which we call Lean-to Hill, I overtook a Ford roadster which had stopped to cool off and water up. The occupants turned out to be Verlin Campbell, owner of Fie Top, and his son Kyle, who lived in town and was driving his father home.

Verlin was a huge man, with a rugged face reminiscent of the cartoon character Alley Oop and a voice that boomed across the gorge beside us and rolled off the ridges. His son Kyle was as tall and his voice as deep, but, more inclined to slenderness, he didn't loom as immense as his father.

No sooner had I made myself known to this enormous man than he exploded with a startling greeting: "Yo're just the man I'm alooking fer! I've got the finest tourist place up here in the world! I want you to see hit!" He waved his hand expansively, taking in the rim of mountains above us. I said that I had come just for that purpose, that I had heard of his place and had always wanted to visit it.

Since the road was dry, we were able to take both cars to the top, Verlin riding with me and enthusiastically describing his place. In a short while, we were in the gap, where the climb ends and the rolling plateau begins. We stopped. To our left, sunk into the slope, was a rough little structure of rock, logs, and shakes, with a roof so irregular and wavy it appeared scalloped. The lower part of the building was of flat rocks, chinked with yellowish mud, while the half-story above was covered with long, wide hand-split shingles. Around this potato cellar leading off vaguely across the grassy slopes were a conglomeration of pole, rail, woven, and barbed-wire fences, patched in spots with weatherbeaten planks. Except for the fragments of wire, the whole layout of building and fences was obviously handmade from material found right on the ground.

A stretch of pole fence meandered back down the hill parallel to the road we had just come up. Across the fence swept a steep, cleared expanse filled with stumps, snags, and log piles, cut by deep gullies. Beyond, there were woods. On the skyline a mile or so away loomed a great mountain timbered halfway up but topped with rolling grassland, rugged, shapely, and appealing. This was Hemphill Bald.

Across a little "laurel" (i.e., rhododendron; in southern Appalachian parlance, rhododendron is "laurel," laurel is "ivy," and ivy doesn't exist) patch in front of us, other cleared land could be discerned, while above the potato cellar was more stump-filled grassland. Still higher, I could see a deadening, the bone-colored tree skeletons silhouetted against the blue sky. Everywhere the soil looked black, the grass and weed growth coarse and heavy.

Though nearly noon on a clear July day, it was cool. A stiff breeze came up the gorge behind us and rustled the leaves of the huge oak trees directly in front of us at the edge of the laurel. Off somewhere in the woods a cow lowed.

We drove around the laurel patch a hundred yards or so and encountered two log cabins twenty feet or so apart. Both had the appearance of age, the logs of one darker than those of the other. One was chinked with concrete, the other with mud and rock. An apple tree stood at the corner of each. The cabin farthest from the laurel patch had a porch, one end of which was boxed into a small room.

This cabin was the home of Verlin and his wife Hattie. The little room on the porch was occupied by a mountain couple, Rufus and Stella Price, who stayed with the Campbells and "helped out." The main part of this cabin consisted of one large room with a fireplace and chimney at one end. In the corner to the left of the fireplace stood a cook stove, the stovepipe going up into the chimney's rock flue.

A table about six feet from the fireplace served both as work surface and dining table. A kitchen cabinet and a couple of other pieces of furniture lined up formed a crude partition through the center of the room. Two double beds, some chairs, and a small table or two were situated in the back half of the room.

Alice Alexander Aumen

Typical of Verlin Campbell's original style of architecture was the three-story, three-room structure shown at upper left that housed his crew of outlaw laborers. Also shown is the bunkhouse being enlarged by Tom as housing for some of his employees. It has since been destroyed.

Across a little branch was another interesting design, a cabin three stories high with but one room to a story. The bottom story of native rock—some laid flat and others on edge—was backed directly into the side of the red clay slope like the potato cellar we had first seen. The other stories, continuing on up into the gables, were of log. This was the bunkhouse. Up the hill about a hundred yards away, sitting on a flat ridge, was a long, low barn, the lower part of field rock and the upper of vertical poles, split in half and nailed side by side, with the bark still on.

Picturesque is too mild a word for these architectural creations. Here was pioneer construction in all its glory. The logs of the three living cabins were flattened and squared like those found throughout the southern Appalachians, notched with the corner dovetails peculiar to these mountains. At this altitude, builders had no access to the wide, easily-hewn, heart-poplar logs so common elsewhere and had to make do with tougher, narrower, more crooked logs of chestnut or oak.

Nearby three squat log buildings served as chicken house, meat house, and wood house. The logs had been left round and were larger than those in the cabins, diminishing in size from the bottom log upward. Their doorways were so low that even a short person had to stoop to enter. There was no chinking between the logs, but long split shakes tacked to the inside overlapped edgeways to form a wall far from weatherproof but good enough to keep out the worst snow and rain.

The roofs of the out-structures fascinated me. The gables were constructed of logs, each shorter than the one below, tapering to a peak. Stretching between these gable logs, other horizontal logs formed the slope of the roof and extended two or three feet beyond the cabins' end walls to provide a wide overhang. Rived boards or shakes were nailed directly to the roof logs. Though unorthodox, this style of roof was attractive and clearly strong enough to support heavy mountain snows.

Only three concessions to modernity were visible in the entire development. The roofs of the home cabin and the great rock barn on the hill were of metal, and the home cabin and rock chimney were chinked and mortared with cement.

I later learned that the main cabin, which was the home of the first settlers on Fie Top, had originally stood across a ridge in an amphitheater-like hollow. As the years went by, their firewood supply had receded from the cabin, necessitating longer and longer sledding for the vital commodity. The pioneers had solved this inconvenience by the simple expedient of dismantling their home and taking it across the hill into the woods, moving the house to the fuel, instead of vice versa.

Propelled no doubt in part by a desire to keep up with the Joneses, that curse of all country landscapes, a tin roof, was put on the cabin. The other modern touch, cement chinking for logs and chimney, required sand, which the builders scooped from a deposit of crumbling rock, and then screened. I have since used this same material for all sorts of cement jobs where structural strength is not vital.

Hattie came in from the field, where she had been working, to fix us dinner, after which Verlin suggested we ride out over the nearby mountains. He saddled up his sorrel mare for me. The little black mule had to carry all 230 pounds of Verlin. Our route took us out through his far fields known as Ned's Licks. "Licks" are spots where settlers call up their cattle—usually every Sunday—after scattering salt on logs and flat rocks. This ritual provides the animals vital salt and an opportunity for farmers to check on sick or injured animals and losses or additions to their herds. Ned's Licks were named for their former owner, Ned Moody, from whom Verlin had acquired the property.

Here was a long stretch of open land blanketing the gap in the ridge and lapping down into the coves on both sides. About five acres were in beans, and two mountain boys were in the patch picking the first ripened crop. The rest of the forty or fifty acres of clearing was in grass except for three scattered one- or two-acre patches of potatoes and a patch of oats. The dirt was black and there were no gullies. Near the center of the cleared land was an almost perfectly round knob, on which we stopped. It was planted in oats, nearly ripe, billowing in the stiff and ever-present breeze. From the crest of the knoll, I saw ridge upon ridge of mountains fading into the pastel blue.

Verlin told me that one year he had raised 935 bushels of produce—potatoes and rutabagas—on one measured acre on top of this knoll. At the time, I discounted this statement. Later after learning what that wonderful soil can do, I no longer doubt it.

Verlin's agricultural plan was to plow up new patches of grassland each year for his potatoes, then re-seed the following spring in meadow grasses and oats. His method of planting was unorthodox. Instead of preparing his ground by plowing, harrowing, and pulverizing before putting in his seeds, he planted as he plowed. Using a hillside plow, he plowed one furrow through the grass on the contour. In this furrow, one man spread fertilizer by hand and another dropped in potato seed pieces. Another furrow was plowed without planting and this covered or partially covered the seed with big chunks of turf. Then one more empty furrow was plowed to space the potato rows before the next furrow was planted. A farm hand followed the plow with a potato rake and pulled turf or dirt over seed pieces not fully covered by the plow. The result was an exceedingly rough-planted hillside field, with great chunks of turf standing on edge or leaning half over. But that turf caught and held the rain, preventing runoff on those steep slopes.

After the plowed fields had been exposed to rain and wind for two to three weeks, dirt and turf settled fairly well around the seed pieces. Then a homemade drag—two-by-six boards five or six feet long, overlapped and nailed together—was dragged over the field. Much of the dirt was knocked loose from the grass roots this way, and after the grass died, a fair job of smoothing down was achieved. Large unbroken clods still remained in the bottom of the furrows, and one might think the tender little potato shoots would have a tough struggle working their way up through them, but this was not the case, and Verlin saved a good deal of time and money by his novel methods of tilling and planting.

The planted fields were cultivated, perhaps twice during the early part of the growing season, with three-footed or five-footed cultivators, and were hoed by hand once or twice thereafter. It was tough hoeing, and usually the fields were not well cleaned.

Verlin relied on two-crop farming. On the last, or "lay-by" working, tender little rutabaga plants were set out by hand between the potato rows. Sometimes these rutabaga plants had been raised in a separate plant bed, but in other cases, seed had been sowed in a light dirt furrow made by a single foot plow down the middle between every second or third potato row. When the plants were large enough, they were thinned to a stand eight to sixteen inches apart, and the surplus were then set out in the blank rows.

The system sounds crude. Planting was haphazard, cultivation careless, and harvesting, which began in October, might extend till Christmas if the freezes had not already ruined the tubers. Yet with the cool and moist weather conditions and the incomparably rich soil, some phenomenal yields were secured. With but little fertilizer, careless seeding, indifferent working, and no spraying, the average per-acre yield was at least three hundred bushels of potatoes and rutabagas together and often much higher.

I was as fascinated by Verlin's baritone descriptions of his methods and successes as by the beautiful soil and sweeping panoramas. It was hard to tear myself away from the fields, but Verlin wanted to take me out over the mountains. We rode on across the fields and into the woods, taking a trail to nearby Hemphill Bald on the Cataloochee Divide. There were a couple of parklike openings along the ridge where sheep were grazing.

"This is the finest sheep country in the world!" said Verlin. "If it warn't fer the dogs, a man could git rich on sheep."

"Can't you protect them against dogs?" I asked.

"No. Nearly everybody in the country keeps a pack of no-count hounds, and they never feed 'em. They jest loafer around over the mountains huntin' somethin' to eat. The worst thing about 'em around sheep is that they don't jest kill one and eat hit. Why, a gang of 'em 'll kill a dozen sheep in a night, jest fer the fun of killin'. You cain't shoot 'em because they work at night, an y' cain't poison 'em without killin' all the good dogs an' game in the country."

Beyond the sheep pasture in the woods, the trail bore left around the slopes of the bald, taking us for several hundred yards through a stretch of young timber and then out into open land near the top of the ridge. The ground here was littered with fallen trees, mostly locust. The weathered trees and logs showed the old circle of girdling axe marks near the butts that denoted deadening. Verlin explained that we were off his land now. Grover Clark owned this huge mountaintop pasture and Grover had contracted the job of deadening to Arch Early some years before. The price for girdling nearly two hundred acres was $2,000, and Arch and his family had lived in a little camp high up on the bald, just below Double Gap, while doing the work. With the help of his family, it had taken him most of two years to do the job. After the girdling was complete, the ground had been sowed with blue grass and clover seed, and a fair catch had been secured. Cattle lay about in the shade of the few scattered live trees. Lower down, however, the deadened areas had already grown back into dense young timber, mostly oak, but with lots of red maple sprouts and seedlings.

Our trail brought us to a rail fence that marked the national park boundary and then to a lookoff point in the open land near the top of the ridge with a surpassing panorama of rolling waves of rugged mountains. The mountain we were on fell sharply away from us. Far below in the gorge could be seen hints of farming land between the ridges in the bottom. Across this deep valley reared the huge, somber-looking Plott Balsams, their height exaggerated by this complete, base-to-crown view and by the deep shadows thrown into the coves by the afternoon sun, now beyond them in the west.

From this angle, the Plott Balsams are the darkest and most rugged-looking mountains of the southern Appalachian region. They are also among the least known. Federal agencies in all their varied acquisition activities have passed them by. The mountains have practically no clearings, for they are too rough and steep for conversion into pastureland. There are no roads nor even trails usable for horses. Bear- and ginseng-hunters are practically the only humans who penetrate the region, and there are places in Campbell Creek and Hornbuckle too rough even for these woodsmen.

Southward and eastward, the distant mountains faded off into the blue of the distance. Across the gorge that formed the head of Fie Branch, along the top of the spur ridge called Fie Top were spread Verlin's fields and cabins. Beyond the fields, the spur terminated abruptly and fell into the deep Jonathans Creek valley. I knew I had found the place.

We were at the high southeastern boundary of the Great Smoky Mountains National Park, on the Cataloochee Divide. We could not look over into the park wilderness, however, because of a wall of virgin oak trees. But Verlin told me that from the top of the bald just above us, we could look over the whole eastern end of the park. We did not make the climb, since Verlin wanted to take me westward along the ridge through the big timber. Our route was an old stock trail that followed the ridge top. The timber varied from big, squat, open-grown oaks on the knobs to tall, slim sugar maples and buckeyes in the gaps, and then to beech and birch, thickest in the steep cove heads. We rode on some two miles beyond Deep Gap to Whim Knob and were back in time for supper.

✣ ✣ ✣

Once I had found it, I had to figure out how to buy it. Some nights I stayed with Verlin and Hattie in their cabin, some I stayed in Asheville, some I spent in half-hearted expeditions to look at sites in other mountain towns. But night and day my mind was on Fie Top. I planned and replanned, dreamed and daydreamed. House here, cabin here, fence here, barn there, water supply here, road there. I saw early on that Verlin was not going to be unreasonable on price, but my funds were limited. I had to think not only about the cost of buying the property but of building and outfitting an entire resort.

It would be an enormous undertaking, and I had qualms about whether we could swing it. I discussed it with my friend, Lush Bramlett, who had furnished horses for our old ranch and was prepared to furnish them for a new one. When I voiced my doubts to him, Lush said, "Oh, go on and do it—Columbus took a chance!" That was the spur I needed.

Verlin thought that he owned about twelve hundred acres. I began with the idea of buying only about a hundred of them, including the buildings and most of the improved farmland. But that wasn't satisfactory to Verlin and Hattie since it would split their remaining holdings. It wouldn't exactly suit me either, because it would be impossible to lay out a hundred acres that would both encompass the buildings and extend to the park boundary. We sat around the fire at night and discussed various propositions.

I called Judy in Virginia and told her to pack up the children and plan to camp on Fie Top for awhile. Miss Hattie and Stella, her hired help, cleaned up the old Moody cabin for us. We brought our own canvas cots, found an old two-eyed, wood-burning stove, a baby bed for Alice, our youngest, and sat around on old potato crates.

Verlin was a remarkable fellow. He had taught school in Cataloochee Valley at one point and during the twenties had operated a store in Maggie, a sawmill community and terminus for a logging railroad. (The empty log-cars were pulled up the surrounding mountains by steam locomotives. Once loaded, the cars were manned with brakemen and turned loose to coast into Maggie under gravity power, often many miles.) The Fies and Moodys, who bought supplies from Verlin, were often in debt to him and had swapped land to settle accounts. He had also acquired some cutover lands from the lumber company to make up his twelve hundred-acre total.

I wanted all of it, but even with help from the bank, I couldn't swing it. Verlin kept the 150-acre amphitheater at the south end of the property, now known as the Verlin pasture, and I bought the rest. A couple of years later, I bought the five hundred-acre Hemphill Bald tract from Grover Clark. This land, reaching 5,700 feet at its highest point, increased our holdings to around fifteen hundred acres in all. I leased part of the

Miss Judy and Verlin Campbell salting sheep on Moody Top in 1938. The Alexanders continued raising sheep for another decade or so, until a combination of dogs and bears eventually made it unprofitable.

Cataloochee sheep approach some haystacks during one of the first winters at the new ranch. At left is the old blacksmith shop, next to the original stable.

bald back to Verlin to grow potatoes and rutabagas. I also bought Verlin's sheep and cattle and made hay to feed them. In those days, we stacked hay instead of baling it, and I learned that there is an art to building a haystack.

As part of the purchase transaction, I bought ten small pigs, which Verlin had kept in the basement of the cabin that Judy, the kids, and I had stayed in. I found out pigs create quite a stench. I struck a deal with a fellow named Hall, who had a good crop of corn, to furnish feed for the pigs in exchange for half the meat in the fall. Some men from Waynesville, who were mainly interested in chitlings, came out to butcher the pigs. They cut their throats and let them bleed to death—pretty awful to watch. Afterward, we cut up the pigs and salted down the meat in Hall's springhouse.

To provide labor for his farm and timber operations, Verlin harbored various Tennessee outlaws, who were safe from prosecution once in North Carolina. One man had killed his father, another had been convicted of bigamy; most of the others were moonshiners. Instead of paying them cash, Verlin fed and housed them in the bunkhouse, sometimes buying them a can of tobacco or a pair of overalls.

As potato king of the county, Verlin raised many hundreds of bushels and kept his outlaw labor busy in the wintertime splitting oak slats and tacking them together to make his cubical crates—about fourteen inches on a side and holding about a bushel. The outlaws had made hundreds, pieces of which were scattered all over the property, hazards to automobile tires. These were the crates we all sat on, after looking them over for protruding nails.

By and large, I got along well with the outlaws, and hired most of them to construct the new ranch. Early on, however, one fugitive, who lived in the first cabin down the road, had noticed me driving up and down the mountain in a car with a Florida license. Late one night, during the period of negotiations with Verlin, this outlaw came up the mountain and woke us up. He asked me to take his wife to the hospital, saying she was about to have a baby. From our days in the Cataloochee Valley, we were accustomed to helping neighbors out, but this man's request somehow struck me as odd. Still, I agreed to go, but nevertheless took along a big flashlight for protection.

On the way down, I asked why he hadn't gone directly to Maggie. He replied that my cabin was closer. When we arrived at his cabin, he asked me to go on in and see how his wife was doing. Feeling more and more that something funny was afoot, I said that was his job. He got out, but instead of heading for his own house, he proceeded to the cabin next door owned by his cousin. I saw both men take a drink from a half-gallon fruit jar. Then the fugitive returned to the car, and I asked about his wife. "Oh, she's all right," he said, so I went back up the mountain. Looking over my shoulder, I could see the two men still drinking.

Next morning while we were sawing firewood with a big crosscut saw, Verlin and some of the outlaws told me that my midnight visitor was a dangerous man. Furthermore, they added, his baby had been born four or five days earlier.

While we were standing there, the man arrived with his cousin and this time asked me to drive him to Tennessee to see his mother, who was very ill. He handed me an envelope and a note, reading, "If you want to see your mother alive, come at once." I noticed that the postmark was two weeks old, so I countered by asking him when his wife had had her baby, letting him know that I knew he had lied the night before. He and his partner turned on their heels and walked off the mountain without another word. I've never known exactly what they had in mind but imagine it was to rob me or kill me for my car.

Right after I hired most of the outlaws, I made the mistake of actually paying them the going wage—ten cents an hour—in cash. Thereupon, every single man went down the mountain, got drunk, and wound up in jail. I bailed all of them out, paying bond equivalent to a week's wages. So the next week, they worked for nothing to repay the money. The third week I paid them cash again, and again they promptly went down and landed in jail.

During construction, we rented a house at Lake Junaluska about twelve miles away, where the oldest of our three youngsters, Tommy, could go to the little mountain school. Branded as a Florida boy, he quickly had to brush up on his fighting and marble-shooting skills.

✣ ✣ ✣

The first thing we needed at the new ranch was a water system. On Moody Top above our proposed development was Indian Spring, named for the Indian called Old Daddy who had lived there. We dug a twelve-by-twelve-by-ten-foot reservoir from which water could flow by gravity to the ranch half a mile away and laid a rock wall inside this hole, pouring concrete in the bottom. But the reservoir leaked so badly that we had to build a wooden form inside it and pour a solid concrete wall. We installed a pipeline to the ranch, digging the three-foot-deep trench with a combination of horse-drawn plow and picks and shovels. I was ably assisted by George Caldwell who had helped me install the plumbing in Cataloochee Valley. The son of a boss mechanic for a lumber company, George was a jack-of-all-trades who knew about electricity, water systems, plumbing, and construction. To make a seven-room house for Verlin, Miss Hattie, and other employees, George helped me enlarge the three-story, three-room bunkhouse by adding two wings of rock and log, one on either side.

Prior to converting the old barn into the main ranch house, we sent measurements to an architect, but we weren't satisfied with his plans, especially his design for a square fireplace. We wanted an enormous, arched fireplace typical of the mountains, with an opening big enough for a man to stand in.

Needing a form to lay the rocks against, we scribed with pencil and string on a big barn door the shape of arch we wanted, then cut the form out and stood it up so we could rest rocks against it. We picked up all the rocks from the fields, including those for the keystone, and hauled them in on homemade sleds behind teams of horses or oxen—hundreds of loads. Rufe Sutton, experienced in building chimneys and fireplaces for mountain cabins, planned the arch, throat, and smoke shelf for this gigantic version. That fireplace has always drawn well and at times has provided the only heat in the big ranch house.

Initially, we installed a coal-fired hot water system, with furnace and cast-iron radiators, to heat the ranch house. This turned out to be one of our less fortunate decisions. The gravel road up the mountain remained impassable throughout much of the winter, so we frequently had to saddle up strings of pack horses to bring up coal to keep the system from freezing and breaking. Even so, we experienced several freeze-ups and broken pipes during the first couple of winters. I finally lost patience and consigned the whole system to the wartime scrap effort. From that time forward, most of the heat in the house was supplied by the great fireplace in the big room.

Fireplace heating differs from central heating in that it warms the body through direct radiation without having much effect on the temperature of the air in the room. On the colder days, the family pretty much lived in the glow of the fire as mountain people did, while in the back of the room we sometimes heard the crack of freezing flower vases.

One winter afternoon, Judy and I were sitting in front of the fireplace, riding out a nasty storm of fog and rain mixed with snow, when Chief Stringfield, Waynesville's police chief, arrived unannounced with some friends, all cold and wet. The chief, an inveterate hunter, had invited some New Jersey friends to hunt bear, but after a couple of days in the Balsams without any luck, the chief decided to bring the men to our place in hopes of better hunting.

A little more than halfway up the mountain, the chief's old Dodge mired up and quit. The men walked up, arriving thoroughly soaked and chilled. I piled several big logs on the great fireplace and brought out a jar of liquor to help the men thaw out. The chief sampled it liberally, while the New Jersey visitors sipped politely.

Guests and Miss Judy around the big fireplace, probably in the fifties. Even in midsummer, a fire is needed every day to break the mountain chill. In wintertime, the fireplace once provided the only heat the Alexander family had.

After his second sample, the chief settled back in his chair beside the fireplace and took a great dip of the black, greasy-looking Copenhagen snuff he favored, then launched into an elaborate hunting story. As he talked, his mouth slowly filled with snuff juice, and it finally became imperative to dispose of it. He turned to the roaring blaze and, with an explosion like a tire blowout, shot a mouthful of the liquid into the fire. He turned back to resume his story, then stopped, looked at the fire once more, faced us again, and finally said, "Now boys, there's a fireplace! Them damned little fireplaces in town, I can spit them out in about two spits, but I'll not spit that'n out."

We had to knock holes in the big barn's rock walls to make openings for windows and doors for the ranch house. The fieldstone walls, two feet thick, had originally been laid up with mud as mortar, and hornets had bored through over the years. One day, shortly after the ranch opened for business, a little garter snake crawled through one of these openings into our new hostess's bedroom, shedding its skin. The sight so terrified the woman that she jumped in her car and left for the city, never to return. Eventually we stopped the holes by covering the mud with concrete on the outside. The mud is still visible inside the house.

We located some cypress at a local building materials dealer, and George Caldwell used it to make windows that lasted forty years before being replaced with newer versions. On our sawmill we slabbed off the sides of round logs to make flat-sided timbers for partitions in the ranch house. To lighten the interior of the big room, we used buckeye,

Tom and his crew raised the roof and knocked holes in the stone walls of Verlin's barn to transform it into a ranch house. Teams of "steers" and horses dragged in rocks and logs that went into the structure. The ranch had neither truck nor tractor until the late forties, depending for the most part on horse-drawn sleds. Tom contended that the invention of the wheel has been overrated: the problem in mountainous terrain is not so much getting a vehicle to move as getting it to stop.

an almost white wood, for the exposed walls. Buckeye rots quickly when exposed to the weather, but these logs would be protected.

In planning a stairway to the second story of the ranch house, originally a hay loft, we made a pattern for George Caldwell to follow in mortising the steps to a central post. Those stairs had to be good and strong, since lots of people would sit on them during square dances in the big room. The floor itself was installed on floor joists of locust logs, resting on rock pillars.

We roofed the ranch house with hand-split oak shingles made on the property—some forty thousand of them—but the pitch of the roof turned out to be too flat and we failed to put roofing paper between the shingles and sheathing, so the roof leaked. After a heavy snow, we would find piles of white stuff in the big room. We lived with this roof for about three years before making a deal with Sears, Roebuck to put on a composition roof. We held a traditional "working" so guests and neighbors could help us remove the shingles; by mountain custom, we provided them a big hot dinner and corn whiskey. While we were building the new ranch house that winter, two bulls, thinking it was still a barn, broke down a temporary door and ranged though the house. The emerging building also provided sanctuary for ewes who gave birth to their lambs in the big room.

To get lumber for our construction projects, I rented a steam-powered sawmill which we hauled to the top of the mountain. We sawed out the old girdled trees scattered across the pastureland, mostly oak and chestnut with some ash and cherry. We also rigged up a dry kiln by tearing out both floors of the old two-story cabin we had been camping in and installing an old steel boiler three feet in diameter and twelve feet long to serve as a furnace. We then stacked the cabin full of green lumber and fired the boiler night and day for months. We dressed the dried lumber on a rented planer, providing lumber for floors, doors, ceilings, and a good deal of the furniture in the ranch house and cabins.

In the early days, we found that many of our urban guests loved turning their hands to ranch chores—milking, haying, trail building, and so forth, and we were happy to see

After sawing trees into boards on their steam-powered sawmills, builders of the new Cataloochee Ranch smoothed them into useful lumber with this gasoline-powered planer.

Encouraged by ranch hand Larry Caldwell (left), guests help saw and split firewood. Like Tom Sawyer, Tom Alexander had a talent for getting people to pay for the privilege of working. Some of the ranch riding trails were pretty much created by guest labor.

them do it. One guest reported that she and a friend were discussing Cataloochee one night at the opera in Atlanta and were overheard by a woman behind, who leaned over and inquired, "Cataloochee Ranch . . . isn't that the place where they make the guests work?"

We pretty much constructed the horse stables Verlin-style. Like some of the more eccentric mountain characters, Verlin was commonly the subject of the anecdotes with which people whiled away the hours at work or over a hunters' fire. Verlin's construction methods generally came up whenever the topic concerned the design of footings for a building. Inevitably some one would stick his thumbs through his overall galluses, rear back, spit, deepen his voice Verlin-fashion, and declare, "Let's just toenail 'er to the ground, boys." Apocryphal or not, just about everyone in the county recognized it as a Verlinism.

In building our stables, we came pretty close to that: we simply planted whole tree trunks into the ground and around them built the framework for the roof, stalls, saddle room, and feed room. It's not a very durable form of construction; after a time the posts rotted off at the ground. At that point, we simply put rock pillars under the solid remainder of the tree trunks and that old horse barn endured for about thirty-five years.

For the first two years, we produced our own electricity with an old Delco generator, installed in Verlin's woodshed. We cooked with wood. Every once in a while the generator would burn out, and I'd have to take the heavy machine off its concrete base and haul it to Asheville to be rewound.

I approached Carolina Power and Light to see if they would wire us up, since they had a line to Maggie three miles below us. They said it would cost $6,000 to bring a line to the top of the mountain. Meanwhile, the Rural Electrification Administration (REA), one of Franklin D. Roosevelt's New Deal projects, had begun bringing electricity to backwoods coves and hollows, enabling people to have lights, washers, stoves, radios, and later, television. REA regulations permitted new lines only if there were at least three users for every mile. Since we were three miles up, I proposed to the local REA board that I would pay the equivalent of nine users if they would bring the line up, and they agreed.

I also talked to the state highway people about improving the road up the mountain. Knowing that we had attracted a lot of tourists at the old ranch, they agreed to bring in a crew from the Works Progress Administration, a Depression-era agency, to improve the road if I furnished a man, a team, and building materials for bridges.

At first, our only telephone service was supplied by a crude static-filled "tree-line"—a single bare wire strung through the trees up the mountain that the state provided after I agreed to become a fire warden for the area. I hoped to tap into a privately owned telephone line running from Waynesville to Maggie. Knowing that Southern Bell was trying to eliminate all the little private systems, I offered to buy the system and applied for a franchise from the Public Service Commission. I sent a copy of the application to the manager of Southern Bell in Asheville, who jumped at the chance to buy the Maggie line and put in a line to the ranch if I would clear the right of way.

We opened for business in May of 1939, having converted two old log cabins (including the dry kiln) and two of Verlin's chicken houses into guest quarters. At first, I employed Verlin as foreman for farming and livestock operations, but that didn't last long. He and Hattie soon built a big stone house in Maggie Valley and moved off the mountain. Our former wrangler at the old ranch, Larry Caldwell, and his wife moved into the bunkhouse to act as foreman, chief wrangler, blacksmith, and jack-of-all-trades.

Horse traders have a saying, "You can't hook him wrong," to describe a horse good at any type of work, and that's an apt description of most mountain men, with whom

One of Verlin Campbell's log chicken houses was paneled inside with new hand-split shingles and converted into this tiny and primitive guest cottage, called "Chigger Pen." It has long since been demolished.

self-sufficiency has always been a powerful tradition. Over the years, we have hired many loyal and capable men, such as Todd Henry, an experienced logger who initially skidded the timber to our little sawmill, then graduated to head guide.

During much of the time I was in Washington or practicing forestry, Glenn Messer capably managed the farm and stable operations. Later, Glenn's son Bill became our resident expert on the care and operation of the expensive new ski lifts we installed in the sixties. Glenn's brother Haynes and his family lived for years in the bunkhouse, doing the milking and looking after livestock. Short and slight, Haynes had a lanky, powerful son, Jake, who also lived in the bunkhouse.

When we first came to Fie Top, we followed the common practice of storing our cured hay in stacks in the fields. Later, I acquired my first piece of self-powered farm machinery in the form of a stationary baler. It had a powerful vertical ram that crammed hay into a central hopper where it got compressed into bales. A reliable and tireless pitchfork hand was required to manually feed good forkfuls into the hopper between strokes of the ram, which moved up and down at about one-second intervals. If the feeder lost his rhythm, the ram would grab and crush his fork, incorporating the fragments into a bale. Deliberate and tireless, Jake Messer was a fine feeder, but the other hands often teased him a bit.

One day, Jake showed up in the hayfield with a brand new pair of work shoes that incorporated steel toecaps designed to protect against crushed toes. Enormously proud of those shoes, Jake spent the rest breaks talking about the strength of those toecaps while his co-workers egged him on. When he speculated that he might even jam that hay baler by sticking in his toe, others said, why, yes, he probably could.

Eventually the other hands went off to bring in a new shock of hay, which they did by encircling it with a log chain and dragging it in behind a workhorse. When they returned to the baler, they noticed that Jake was standing off looking quiet and pale, and then they saw that one of his new shoes looked mangled. They guessed what had happened, and Jake confirmed that he had indeed tried jamming the baler with a shoe. Asked if he was hurt, Jake said he might have sprained his toe a little and began feeding the baler.

Then the men noticed that blood was running out of the tears in Jake's shoe. They got him onto a hay sled and drove it to the ranch house to let Judy take a look. Over the years, she has provided first aid to many a bloodied worker or guest—splinting compound fractures, stitching axe wounds, bandaging legs mangled in bear traps—but she said that taking off Jake's boot that day was almost more than she could take. When it finally came off, the big toe was found to be completely severed from the foot. She stanched the bleeding as best she could, got Jake into the station wagon, and headed for the Haywood County Hospital in Waynesville, fifteen miles away.

After she had left, Jake's father Haynes fished the toe out of the boot and carried it out to the hayfield. When the others had looked at it, dismayed and chagrined, little Haynes wandered off to the fence corner and stood looking off for a minute. Finally, he was heard to mutter, "Ah . . . hell," as he tossed the toe into the weeds.

Mr. Tom and ranch hands in the forties. From left to right are Frank Howell, Haynes Messer, Mr. Tom, Shuford Green, Glenn Messer, Jake Messer, and Broady Messer.

Early on, we acquired the priceless services of Shuford Green [*who stayed until his death in 1991*—ed.]. A tireless and dedicated worker, Shuford walked to work every morning and back home every evening, a round-trip distance of about six miles. The other men often commented that if a new man could keep up with Shoof for a day, he would be a good man to hire. The only time anyone can remember Shuford missing a day was late in his life when he decided to get married. That day, he told me he had to leave at noon, but he was back at work the next morning.

But Shuford also had definite ideas about the fitness of things. When the government instituted "war time," or daylight savings time, during World War II, the entire ranch had to remain on the old peacetime schedule because Shoof insisted on coming in at the "real" time. Not long afterward, he was drafted into the army despite my pleas to the draft board that he be exempted on the grounds that as a working farm, we contributed to the war effort. A few weeks after Shoof went off to boot camp, I got a call from his commanding officer wondering whether I would take him back and put him to work. The officer complained that Shuford simply refused to obey many orders; for instance, he wouldn't pick up cigarette butts from the training areas on the grounds that he "hadn't throwed them there." And he refused to wear socks since he had never in his life worn socks in the summer time.

In December of 1961, when we first started the ski operation, I asked Shoof to be part of the snow-making team. He reluctantly agreed, but at 8:00 p.m., as the weather was just getting cold enough to turn on the snow guns, he picked up his dinner bucket and left the mountain, saying he had never spent a night away from home and couldn't start now. Nevertheless, Shoof made a fine and well-loved daytime ski-lift attendant for many years afterward.

The rate of pay, originally ten cents an hour, has gradually risen with the times. Our employees have always been outstandingly honest in regard to wages. If I overpay a man, he lets me know as quickly as he does when I underpay him.

On a typical day, the men meet at the stable at 7:00 a.m. to get their orders to collect firewood for cabins and ranch, check and repair fences, clear trails, plow fields, bale hay, or repair water lines, whatever the day requires. The men round up the cattle, move them from one pasture to another, cut calves from cows and wean them, feed steers for butchering time, feed horses. Our horses roam about fifty acres of pastureland, coming to the barn early in the morning to be fed.

Soon after we opened, two former guests from old Cataloochee, Edith Begni of Paris, Texas, and Irma Bain of Lexington, Kentucky, bought small tracts of our land. I designed and our crew hewed and erected attractive log and stone houses for them, providing us and them winter income. Over the years, we sold parcels and lots to other friends and former guests, including Bettina and George Shackelford, Babs and Arthur Chenoweth, Charlotte and Raymer Maguire, Jean and Rick Carter, and Sue and George Brownell.

Horseback riding and hiking have always been the principal diversions at the ranch, with pack trips of various lengths available for the most dedicated riders and campers. We continued our old relationship with the American Forestry Association, which sponsors pack trips in many wilderness areas in the United States. We generally conduct two ten-day trips into the Smokies each year, one in the spring and one in fall. For many years, pack horses carried all the supplies for the ten-day trips. A supply truck would meet the riders once, at the halfway point.

Looking west from the present ranch toward the Plott Balsam range and Waterrock Knob.

N.C. Department of Conservation and Development-Cataloochee Ranch Collection

One of the problems of pack tripping in the lushly forested eastern United States is that comparatively little natural grazing is available. Virtually all the horse feed has to be packed along. That, of course, takes lots of pack horses, therefore more feed, therefore still more pack horses, etc. Even with a midpoint supply, a typical group of twenty riders on a ten-day trip will be accompanied by a pack string of twenty horses. At best, leading and managing that many animals takes a lot of manpower and at worst, it can become a nightmare. Yellow-jacket season is generally the worst.

From the middle of summer through the first cold days of fall, jackets build their nests in holes in the ground in sunny spots, very frequently in the middle of trails. Generally the jackets don't sting the riders, nor even the first two or three horses in a string. It's the later horses that bear the wrath of the stirred-up bees. We have learned that the best way to get through a nest is for the rider who spots it to yell, "jackets!" signaling riders in front to spur on their horses and riders in back to hold back until there is enough room for them to gallop through the nests. With this system, most horses get by unscathed, though we have had a few tormented animals jump off the trail and into the bushes trying to brush off the bees.

The pack string, which generally travels ahead of the main party of riders, often experiences the worst problems. Each wrangler usually leads four or five pack animals, tied one behind the other. When they get into the middle of a swarm of jackets, hell can break loose, with heavily laden horses rearing and flying off the trail down the mountainside, occasionally winding up upside down on top of their packs in heavy brush or a creek, unable to get back on their feet.

Jackets often pick inconvenient places to build their nests, like the stretch of Appalachian Trail called the Saw Teeth, on the crest of the Smokies along the state line. Much of that trail was built by the CCC and others atop narrow rock walls along the sharp ridge, with sheer drops into Tennessee on one side and North Carolina on the other. On one trip, a twisting, jacket-covered pack mare stepped off this wall, leaving her forequarters in the trail but hind legs flapping over empty space. Swatting off bees, the wranglers managed to get the frantic animal back on the trail, only to discover she had severed a major leg tendon on the jagged rocks. This was some thirty miles from the ranch and ten miles from the nearest road.

The wranglers sat down to wait for me, leading the main party, to arrive. Normally, a badly crippled horse would have to dispatched, but as always on major pack trips, we had a physician along. After examining the animal, he volunteered to attempt to repair the leg. With lots of sweat and rope, we immobilized the mare atop the wall, and with plenty of suture material, the doctor stitched the tendon ends together. Miraculously, the animal managed to limp as far as the highway, where we could phone the ranch for a truck to take her home, and, ultimately, she recovered completely.

The guests on one trip included a rather hefty lady who had trouble getting on and off her horse. At one lunch break, as she slid down the side of her horse, her bra caught over the saddle horn; there she hung, feet not even on the ground. She yelled to wrangler

Todd Henry to help her undo the bra. Todd came over, surveyed the situation, and said, "Madam, I'm not too familiar with these things. Did you bring a blueprint with you?"

"Blueprint, hell! There's just three hooks." But after struggling a good while with the straining fasteners, Todd finally had to pull out his pocketknife and cut her down.

Todd also took some teenage boys to Spruce Mountain. One spoiled kid wouldn't obey Todd's orders and believed nothing he was told. Asked to stop picking the wildflowers, he ignored the directive. Told there were bears in the Smokies, he scoffed.

While gathering firewood at the Spruce Mountain campground, Todd spotted a young bear that seemed pretty tame, having fed on enough camp scraps to lose his fear of mankind. After supper and tale-telling was over, the boys retired to their sleeping bags. When they were asleep, Todd found some honey and spread it over the foot of the troublesome kid's sleeping bag. Eventually the little bear, attracted by the smell, came along and started licking up the honey.

Todd yelled at the animal, figuring it would try to pick up its goodies and flee. Sure enough, the bear grabbed the foot of the sleeping bag and headed down through the woods, dragging boy and all. The boy hollered so loud that eventually the bear dropped his burden and fled. Todd came down, helped the boy unzip the bag, and asked him whether he thought there were bears in the Smokies.

"Yes, sir, I sure do!" said the kid and gave no trouble thereafter.

On one of the big American Forestry Association trips, the party was camped at Walnut Bottom, which has a small spring with wonderfully cold, clear water. Sam Woody, the cook, had installed a short length of pipe in the spring to make a spout of running water and had also hung a pot at the end of the pipe in which to chill butter. During the night a wandering bear came through, smelled the butter, and stuck his head into the pot. When he lifted his head, the pot bale lodged behind his ears, with the pot covering his face and eyes. Blind and scared, the bear went running in every direction, blundering into trees and tents. Horses were breaking loose, the whole camp yelling–while Sam was chasing around behind the bear, concerned only about his pot and butter.

With so many people and animals out for protracted periods in strange and remote surroundings, I've always been amazed by how few really dangerous episodes and serious injuries we've experienced on pack trips over the years. Horses, by and large, are intelligent and trail-wise animals, craving human care and companionship. But at the same time, they have flighty, distractible tendencies that can easily get them into a cascade of trouble without a firm human hand. I have always tried to indoctrinate both crew and customers with the importance of controlling their horses at all times.

On one ten-day trip that I wasn't along on, about sixteen riders left Cabin Flats on Bradley Fork, climbed over Hughes Ridge, and followed the magnificent wild trail down to Raven Fork by way of Enloe Creek. They lunched at Raven Fork, marveling at the wild beauty of the stream, where the force of the water squeezing between enormous boulders creates a roar that can be mistaken for thunder.

N.C. State News Bureau–Cataloochee Ranch Collection

Town and Country Studio—Cataloochee Ranch Collection

Facing page, top, guests and horses shortly after the new ranch opened in 1939. At this time the ranch house still had a roof of hand-split shingles, which were removed soon afterward because they leaked. Middle, a pack string departing. Bottom, riders setting out from the old stables—the ones constructed Verlin-fashion. Miss Judy's vegetable garden lies behind the riders. Above, riders head through Ned's Licks.

Lunch over, head wrangler Todd Henry hollered, "Mount up!" and spent a few minutes instructing riders about crossing the surging stream and keeping horses away from the hidden underwater boulders. Other wranglers were spotted throughout the string, with daughter Juju fifth in line and her son, fourteen-year-old Richard, riding "drag," or last.

The crossing started out fine, with everyone obeying directives, pulling up their feet, guiding horses properly, and splashing out the other side. But the wife of a New York dentist, who was not really enjoying the wilds very much, let her mare Rosie wander in the middle of the stream. Rosie fell over a big rock and they both went for a swim. The dentist yelled and the string stopped while the woman waded out. Then he insisted that a halt be called and a tent of ponchos be constructed so his wife could change clothes.

This was not really a very good place to halt a string of horses. As the narrow trail emerges from the creek, it grades up Hyatt Ridge above an almost vertical drop down a rock face to the bed of Raven Fork, where boulders are scattered like enormous marbles. In line behind Todd was the horse Flint, carrying a gentle and timid rider, followed by Lady Bug. As soon as the horses stopped, they naturally tried to bend their heads for bits of grass—a trait that, unless forestalled, often gets them and riders into trouble. As Flint nibbled grass on the trail bank, he stepped backward and placed a hind foot inside the halter rope of Lady Bug. The tangled horses tried to extricate themselves. The woman on Flint became frightened and pulled back on the reins, causing Flint to rear and back up. Todd yelled for her to jump, which she quickly and pluckily did, while Flint continued over the edge and slid upside down and head first into the creek, winding up between two huge rocks that leaned together almost like a tent. He regained his feet and didn't seem hurt.

Having assured themselves that the riders were all right, the wranglers concentrated on getting Flint back onto the trail. A horse's steel shoes give no purchase on slick rock, and five or six people pulling on Flint's reins and halter did no more than make him resist their efforts.

As time went on, some of the riders grew restless, some even wondering how long they would have to stay there before leaving the animal. Eventually, Todd decided to take part of the group to camp at Round Bottom, six miles further on. From there he planned to send a smaller party on to the ranch, some twenty miles beyond that, for tackle and slings to rescue Flint.

Those who remained sat down to think. Juju had already gone down and removed Flint's saddle and bridle, and as she was scrambling back up, she noticed a small ledge, about four feet by four feet, over on the right. They speculated that with two jumps over rocks, Flint might reach that ledge—if he would. From there it might be possible to construct a bridge back to the trail. In a short time, young Richard had used the axe to fell a tall, ten-inch poplar, and recut it into two poles long enough for the bridge. These poles were installed between the ledge and the trail, leaving about an eight-inch space between the poles that could be covered with large flat rocks.

While riders searched out rocks in the creek, Juju went down to Flint, took him by the halter and talked to him, trying to persuade him that he could make it to that little ledge. She is convinced Flint tried to listen and understand that all the activity was on his behalf. Finally, taking a firm hold on his halter, she got him to scramble to the ledge. Very, very delicately, Flint turned around in place so he could watch the rock-carrying activity. Clearly comprehending what was going on, he began prancing with his front feet, and Juju even had to restrain him from crossing before the bridge was completed. The bridge was narrow and Juju knew that Flint was going to be in a hurry, so she'd better not be in the way. As the last rock was put into place, she unsnapped the halter rope and turned him loose. He had to make about an eighteen-inch step up onto the bridge, which he did carefully, and then very quickly made the trip to the trail. Tears came into everyone's eyes as they heard every horse present nicker, "Welcome and well done!"

They got into camp in time to stop the other rescue party and to have a celebrating drink before bed.

After sixty-odd years of pack tripping in the Smokies, Cataloochee wranglers feel about as comfortable in camp as they do at home. Paying guests generally pitch in on chores, and most everybody is ready for bed early in the evening.

Jack Szold—Cataloochee Ranch Collection

Typical of the old upland homesteads was the cabin of Uncle Charlie Rich of Rich Cove, just out the ridge from Fie Top. The Rich family did a little subsistence farming and raised some livestock. Much of its livelihood came from the surrounding woods in the form of groundhogs, squirrels, and bear, ginseng root, and the occasional contribution from whiskey distilling. When developers planed off the top of Buck Mountain to establish the Ghost Town theme park above Maggie, they demolished the cabin.

When we first arrived at Fie Top in the late thirties, some of our mountain neighbors still survived partly by hunting, trapping, and gathering wild materials such as chestnuts to eat and ginseng roots to sell. Several times a year, we might see, for instance, the men from the Rich family, who lived in a neighboring cove, heading across our fields for the woods with rifles on their shoulders and a dog or two in tow. Their sole camping gear would consist of hunting jackets whose pockets might contain a length of fishing line and some hooks, a chunk or two of corn bread and perhaps a little meal, salt, and ground coffee. Thus equipped, they could remain in the wilds for days, rain or shine, sleeping on the bare ground under an overhanging rock or between two big fires, subsisting largely on what they killed, caught, picked, or dug up—squirrels, groundhogs, trout, grouse, nuts, berries, wild leeks, and so forth.

Mostly their armament would consist of .22-caliber rifles or single-shot shotguns, but sometimes one of the older men would be seen carrying an ancient muzzle-loading rifle, the long, graceful weapon of the type used by Daniel Boone and the frontier fighters of

Probably before a muzzle-loading rifle shooting match at Cataloochee Ranch in the mid-forties, Tom Alexander (center) chats with Uncle Charlie Rich (left) and his brother, Frank Rich (right).

the American Revolution. These shoot homemade round lead balls, much larger than a .22 bullet, and possess greater killing power for dispatching large animals such as bear or deer. Compared to a modern high-powered rifle cartridge, the lead and black-powder ammunition costs almost nothing, so muzzle-loaders were frequently used to shoot pigs at hog-killing time and are still known throughout the mountains as "hog rifles." Most old families still had one or two, however rusted or in disrepair.

While the oldest muzzle-loaders were originally built as flint-locks, virtually all were long ago converted to percussion-locks; that is, the powder-charge is no longer ignited by sparks from a flint but by the impact of the hammer on a percussion cap, a tiny copper cup containing a volatile explosive. Though a surer-fire means of ignition than flints, percussion caps occasionally disintegrate in the shooter's face. One neighbor down the road, Wesley Gibson, was blinded in one eye thereby. During a hog-killing time, Wesley was shooting hogs in a pen when a percussion cap disintegrated, throwing fragments into his left eye. Seeing blood streaming down his face, Wesley's companion urged him to get to the hospital. "Hit's too late fer that; that eye's gone," Wesley replied. "We might as well finish killing these hogs."

At the end of a long, hot haying season in the mid-forties, I suggested to the ten-man work crew that we ought to pitch some kind of celebration. Talking it over, one of the older men suggested an old-time muzzle-loading rifle shooting match. The suggestion was enthusiastically taken up by the other men. Shooting matches were an ancient mountain sporting event that was almost extinct by then, but some of the older hands remembered the rules. Contestants were usually divided into four age classes—under twenty years, twenty to forty, forty to sixty, and over sixty. Prizes consisted of quarters of a steer,

Contestant Charles Ray of Waynesville takes aim from a bearskin pad at an early muzzle-loading rifle match held at Cataloochee Ranch. Contestants had their choice of firing from two positions, either prone at a distance of sixty yards from the target or "off-hand" (standing) from forty yards. Most preferred the prone position to trying to hold steady these long and heavy Kentucky-style rifles, many of which were, and still are, highly accurate at considerable distances. Rifled hunting firearms similar to these were the "secret weapons" that enabled American frontiersmen to defeat their more numerous enemies in many battles of the American Revolution.

N.C. State News Bureau—Cataloochee Ranch Collection

Rufe Sutton, stonemason and bear hunter, loads up. Care in loading, which includes a uniform powder charge and pressure in seating the round bullet, is a major contributor to accurate shooting.

with the forequarters going to the younger two classes and hindquarters to the older two. The overall best shot won the hide and tallow as well.

Contestants paid an entry fee of one dollar and either brought their own targets or obtained the makings from the ranch. The targets consisted of a short piece of charred board over which was tacked a sheet of paper with an aiming point—usually a diamond-shaped hole—cut in its middle. Somewhere on that board, there had to be marked a cross consisting of two knife cuts. Each contestant presumably knew his rifle's idiosyncrasies

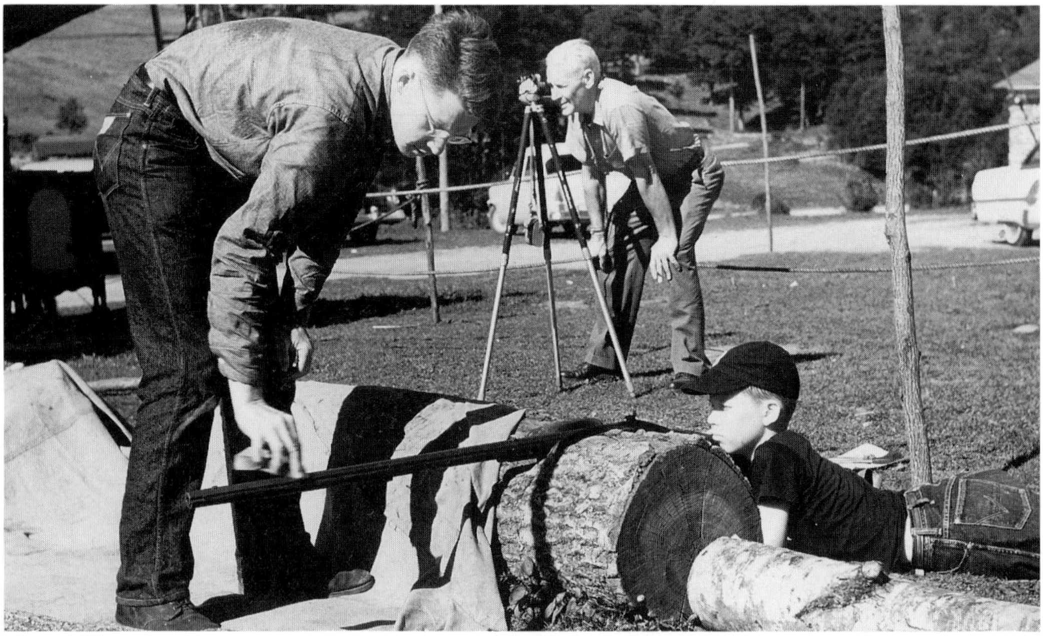

G. Martin–N.C. State News Bureau–Cataloochee Ranch Collection

George Alexander takes aim, while brother Tommy shades his sights, and big Tom spots the target.

as well as its "zero" (where the gun will actually place its bullets on the target), and so each could put the cross wherever he desired. Each contestant fired once in succession. Once all had fired three times, the judges would remove the paper from the board and carefully measure each bullet hole's distance from the cross, totaling the distances for the three shots. The winners, of course, are the ones with the lowest totals.

That first year, only our men and a few neighbors entered the match, but the following year we opened the contests to anyone who wanted to enter. In time, as many as 150 contestants attended, some from far-off states. We even started a special women's class that attracted some modern Annie Oakleys who held their own against the best of the men. As the matches attracted more and more expert marksmen, many bringing fancy equipment and telescopic sights, they also brought different expectations about how proper shooting matches should be conducted. By the late sixties, the clash of cultures generated so much ill will and controversy that I reluctantly concluded we had to give them up.

✣ ✣ ✣

To supplement their meager incomes in those hardscrabble days, many of the local men made whiskey, an honorable profession requiring hard work and honesty to keep the product pure and safe. Stills were hidden throughout the mountains, usually in dense thickets near a good source of water. Only people carrying a grudge snitched to the lawmen about the locations. I knew where most of the nearby stills were and even had arrangements with some of the moonshiners to provide me a supply, but I tried to insist that they keep their stills off ranch property.

Somewhat halfheartedly, Tom tried to discourage whiskey-making operations on ranch property. But every so often a hunter would stumble across a still hidden in a laurel slick not far from water. The operator usually turned out to be a neighbor. Tom couldn't report the man to the law, of course, but he sometimes negotiated a small toll in the form of product or demonstration for ranch guests.

Even so, I remember looking out the window early one morning in late November after a windy night had stripped the trees of leaves and spotting the glow of a still fire on a nearby hillside on ranch property. Naturally, if I could see the whiskey makers' fire, they could see the ranch house lights. The next day, there was no fire.

On another occasion, I stumbled on another still in the middle of the ranch property, hardly a mile from the house. Sensing an opportunity for entertainment, I first talked things over with the moonshiners, who were well known to me. Then, at breakfast the next morning, I told the guests that they could see an authentic mountain still in operation and try a taste of the real thing if they would meet me at the barn. After the guests had mounted horses, we blindfolded them and set off on a long, circuitous journey, finally dismounting at the still. Everyone had a fine time. The moonshiners enjoyed showing off their rigs and made a little extra income selling samples to the guests. The guests, in turn, were delighted at the opportunity to experience authentic white lightnin' at its source.

✣ ✣ ✣

I had always viewed it as essential that we continue real farming and grazing operations on the ranch. Having so much black, well-watered, disease-free mountain land, we expanded Verlin's potato raising into a certified-seed business. Raising seed certified to be disease-free requires frequent state inspections of each year's growing crop and harvested potatoes, discarding diseased plants and potatoes, as well as crop rotation so that potatoes don't get planted in the same fields two years in a row. With many acres of fine open land available, that was never a problem.

Raisers of certified seed generally acquire their own seed from distant regions to lessen the likelihood that diseases will reproduce. We frequently exchanged potatoes with regions as far away as Maine—or thought we did. One fall, while sacking up potatoes for shipment to the local seed broker, I lost the fancy pocketknife I had been using to cut tie-strings. Next spring, we bought seed potatoes from that same broker, who naturally got a commission on both transactions. When I opened one of the first sacks of "Maine" potatoes, out fell my knife.

The steepness and roughness of the terrain doesn't lend itself to farm mechanization, so southern Appalachian people have been among the last Americans to abandon animal-drawn logging, plowing, sledding, and other remnants of nineteenth-century technology. Powerful, slow-moving ox teams, or "steers" as they are known locally, dragged in many of the logs and rocks that went into the construction of the first ranch buildings. All of our potato crops and most of our hay and garden produce were cultivated and harvested with the aid of horses or mules.

When you harness up a horse in the morning and proceed to the field, you never

N.C. Department of Conservation and Development—Cataloochee Ranch Collection

During his stewardship, Tom Alexander deeply wanted to see Cataloochee's extensive land holdings pay their way. Mainly he worried about ranch employees being dependent for their livelihood on the seasonal tourist business. He experimented constantly—raising livestock, row-crops, timbering, even shrub and moss cultivation. Like many another landowner, however, he found that rising wages and flat-country competition made mountain agriculture an increasingly marginal activity. Now, direct economic returns to the ranch from its extensive acreage are mainly limited to cattle grazing and some timbering.

When bears left them alone, cattle were a substantial source of cash in the early days on Fie Top. Tom Alexander introduced some of the first Aberdeen-Angus cattle into Haywood County.

really know how the day is going to unfold, but it's not unlikely that at some point the tedium will be relieved by the spectacle of a mowing machine, hay rake, or sled bounding down a mountainside, rapidly disassembling itself behind a runaway team spooked by yellow jackets, lightning, or equine imagination.

As part of the deal with Verlin Campbell, we acquired his substantial sheep and cattle herds and therefore continued the summertime haying operations essential to carrying the animals through the winters. In the forties, I began replacing Verlin's cattle, which were mainly mixed breeds, with purebred black Angus, then new to this region. The Angus thrived in these mountains, whose climate so resembles that of the Scottish highlands where the breed was developed.

We expected and encountered the innumerable afflictions that go with raising livestock. Sheep are especially prone to troubles, such as difficult births in hilltop blizzards, leaving orphans that have to be raised on bottles. We learned how it is often possible to induce a ewe that has lost her lamb to adopt and nurse an orphan by tying the familiar-smelling skin of her dead infant over the orphan's own.

What we hadn't counted on were the predators—dogs and bears. Mountain hunters generally keep their hounds lean and hungry to maintain hunting enthusiasm, and often they don't keep them penned, or else the dogs simply get lost during a hunt and turn wild. On many nights we would hear the cries of panicky sheep on the hillside and go out to find many slaughtered, apparently out of sheer fun, by packs of dogs. One year I lost forty-five sheep to dogs alone. We also shot some dogs, but that can be a dangerous remedy, since many hunters, like fond parents, can't believe their charges capable of criminal behavior. One day I ran upon a pack of dogs that were still feeding upon the remains of two sheep. I shot two of them before they could get away. It wasn't long before some neighbors called on me. "Tom, did you shoot our dogs?" one asked. "We want the facts."

I said, "I won't lie to you. They were in my sheep when I shot them. Do you want their collars back?"

One man replied, "We don't blame you, Tom," and the others nodded in agreement. They're now among my very best friends.

Worse than the dogs, however, were bears. The black bear is as American as corn. He has lived all over the continent from the Atlantic to the Pacific and from the Gulf of Mexico to the Arctic Ocean. The first white pioneers ran into him wherever they settled and soon discovered what a rogue he was, robbing beehives, cornfields, chicken houses, pig pens, and sheep and cattle yards. The pressure of increasing human population eventually eliminated the bear from broad areas of the country and pressed him into sanctuaries. In eastern America, these sanctuaries include the rugged mountains of the northern and southern Appalachian chain, dense coastal swamps, and inland river bottoms.

In the mountains, his numbers were kept in check by hunters before state and federal laws—and particularly the Great Smoky Mountains National Park—began protecting him. Always a favorite sport, bear hunting became even more popular as park protection increased the bear population. To get enough food, the bears are often forced to forage outside the park in apple orchards, cornfields, and livestock ranges.

Like man-eating tigers, only certain bears become killers. To this day, in fact, some so-called wildlife experts flatly maintain that black bears are vegetarians, confining their meat-eating habits to excavated grub worms or bee larvae. Presumably these experts have never witnessed the savage aftermath of the clash between a bear and a grown cow on a thickety mountainside—the broken and uprooted bushes marking the route of the frantic cow's attempted escape, the plowed earth of the kill scene, the clawed sides and broken neck of the victim. Often the bear eats only the udder that first night but comes back on many subsequent nights to dine and wallow on the decaying carcass.

Our bear troubles began early one morning the spring we opened the new ranch. Rufe Sutton, a neighbor who often helped us out, was at the barn when he heard some nearby sheep bleating. He struck out up the hill and within three hundred yards came upon a bear nearly as big as a bull standing between the carcass of one dead sheep and the dying body of another. That night we set a trap near the dead sheep and caught the bear. Before the bears retired that fall, they had killed forty-two of our sheep and seven beef cattle. Meanwhile, though, we caught several bears.

Over the years, I've become convinced that the black bear has to be one of the most intelligent animals on four legs. I'm sure, for instance, that many somehow learn that the national park fence represents the boundary between danger and safety; otherwise, how can one account for the fact that some bears will kill an animal on private land and then carry or drag its heavy carcass uphill for a quarter of a mile, hoist it over the park's rail fence, and drag it a few hundred yards into a park thicket before sitting down to dine?

After losing several sheep to the nocturnal marauders, we moved the flock to pastures farthest from the park and put cattle on Hemphill Bald. That didn't solve the problem; not only did the bears continue killing sheep, they also began killing cattle.

Outside the park, bear hunting was legal in certain seasons, usually fall. I had many discussions with state game authorities about how to protect my stock. According to law, I could not trap or poison predators at any time, though I could shoot them if I caught them in the act of killing. But these stock-killing bears only operated in darkness and generally only in pastures far from my house, so it was almost impossible to be on hand when the bear attacked. I explained this time and again to game authorities and asked

Traditionally, bear hunting has held a romantic appeal for mountain men unequalled by anything else except perhaps liquor making. At Cataloochee it wasn't so much a matter of romance as necessity: marauding bears from the adjoining national park constantly threatened the economic viability of livestock raising. A lot of sheep and cattle got killed but a lot of bears were hunted down or trapped—and not always entirely legally. A typical homemade bear trap like the one shown below weighs between seventy-five and a hundred pounds.

S. H. Essary—Courtesy of Hunter Library, Western Carolina University

for relief in two ways: first, permission to trap killer bears at the carcass of the dead stock, and second, to employ a capable woodsman who would patrol the park borders, track stock-killing bears, and organize hunts to eliminate predators.

Eventually, the authorities bent the law to this extent: they told me I could set traps if I would then keep alive any animals caught and turn them over to the wardens, who would transport the animals to distant woods and release them. That's what we proceeded to do.

The first opportunity came in June of 1940, when we found a fine, freshly killed young heifer above the Dripping Spring on the left-hand side of the bald. I called in some neighbors known to be expert in the art of trapping bears.

A bear trap is nothing more than a very large steel trap. Many are handmade by mountain blacksmiths and may weigh seventy-five pounds or more. Taking advantage of the propensity of killer bears to revisit the scene of the crime, trappers build a light two-sided pen of poles around the carcass and set their traps in the opening, covering them with leaves.

An unusual feature of bear trapping is that the trap is generally not fastened down. Though the jaws are equipped with inch-long teeth and are activated by powerful springs that can only be compressed with the aid of clamps or long pole-levers, a bear of any size can generally pull its foot free of a freshly sprung trap unless the trap is free to move. So every bear trap comes equipped with a three-pronged grapnel hook which is attached to the trap by a short length of chain.

When the bear springs the trap and heads off through the woods toward a laurel or rhododendron thicket, the grapnel drags behind, catching on the underbrush, slowing him down and fatiguing him and allowing the teeth to work their way in and get a firmer hold. The trail is generally pretty easy to follow. Usually the animal travels no more than half a mile or so before becoming inextricably tangled in brush.

So, early on that fine June morning, Rufe Sutton came in to report that we had a bear in a trap. I assembled a bear-tying crew, consisting of all the ranch hands, and notified guests that they were welcome to join the project. We headed for the scene, deep in the woods a mile and a half away.

Tying a live bear is one of the more interesting ways I know of to spend a morning. When we came upon the scene, deep in a rhododendron thicket below the bald, we found the animal, a medium-sized male weighing about three hundred pounds, in a fifteen-foot circle of black loamy ground that looked as though it had been freshly cleared and plowed for planting. A rank primitive smell of fatigue and desperation surrounded the trapped and powerful creature, which continued lunging against the chain wrapped around the base of a frayed clump of rhododendron and caught by the hook. I noted that the clump could be uprooted at any time and, fearful of what might then befall the encircling spectators, made sure that at least two men kept high-powered rifles trained on the bear.

After consultation, we agreed that the first order of business was to immobilize the powerful snapping jaws. My ten-year-old son Tommy, the resident lariat expert, was designated to throw two loops of rope over the bear's head. Then teams of men pulled on the ropes from either side, while one brave soul slipped up behind, placed a large stick between the animal's jaws and tied it in place behind its head. We used similar tactics to immobilize and bind together the enormously muscled front legs, capable of disemboweling a horse with a single slashing blow. Finally, after about three hours of work by half a dozen men, we also had the hind legs bound, a stout pole between the pairs of legs and the head supported from the pole by ropes. Three men got on each end of the pole and we started out of the woods with the live bear.

For reasons I've never understood, he didn't survive the trip. Whether from fatigue and stress and ultimate heart failure or from the unaccustomed upside-down position under the pole, he simply expired before we could get him to the nearest road, about a mile away. I notified the game wardens who came up and examined the animal. They found no external signs of damage other than the trap-damaged paw, but they nevertheless insisted that we continue tying any bears we trapped in the future. Over the next year or so we trapped and tied two more stock killers, both of which died on the pole before we could turn them over to the wardens.

As years went on, we lost more and more valuable sheep and cattle, always in the summer and fall before the official bear-hunting season began. One result was that we quit raising sheep altogether. Another was that we sometimes simply declared our own hunting season in the sovereign state of Cataloochee.

The typical bear hunt would begin after I heard a cow bawling for a missing calf or spotted flocks of buzzards circling over the face of the bald. Hiking to the area, I would generally manage to strike bear tracks and follow them to the scene of a kill. Sometimes we simply set illegal traps, other times I called in my bear-hunting neighbors, who never failed to respond, in season or out.

Bear hunting is an enterprise in which expert hunters split into two main groups. The smaller group consists of drivers and their dogs, the larger consists of "standers," individual hunters stationed in roads or strategic gaps in the ridges where bears might cross, perhaps to their sanctuary in the park. The driver with the best strike dog picks up the fresh track and determines which direction the bear is heading. Then the other drivers turn their dogs loose and listen to the music as they take off on the bear's trail.

The mounting clamor as the pack approaches his stand has to be one of the most exciting, anxiety-producing events a hunter can experience. I won't forget the evening I was stationed on a spur ridge that leads off the bald. The dogs had struck the bear's trail shortly before sundown in the Devil's Britches, a densely thicketed hollow far below, and for the last hour I had heard the pack making its way back and forth along the mountain face, intermittently falling into silence as it lost the trail, then picking up the chorus as it found it again, but always coming closer and closer while the woods grew darker and darker. I nervously paced up and down my patch of ridge, trying to guess where the bear would cross.

Courtesy of Hunter Library, Western Carolina University

Old pictures of Appalachian hunters, bears, and traps dating to early in this century show that not much was different from Tom's description.

Frank W. Bicknell–Courtesy of the N.C. Division of Archives and History

Then, just as the darkness was almost total, I made out the black shape loping through the little ridgetop clearing, not fifty feet away. I raised my Spanish-American War–vintage Krag-Jorgensen rifle, but could no longer make out the sights. Sighting as best I could, I fired—and the bear kept going. I worked the bolt and fired again—and the bear kept going, crossed into the bushes and was gone, headed for the Yaller Patch, Pinch Gut, and safety. Mortified, I made my way down to the next gap, where some of the other hunters had already gathered and were building a little fire. By ancient custom, a hunter who misses loses his shirttail; I entered the firelight, pulling out my shirt. "Cut 'er off, boys." I said. And they did. Forever since, the scene of my failure has been known as Tom's Stand.

Often when the bear gets through the ring of standers, the hounds, followed by their drivers, catch up with the animal where he's resting. A terrible fight often ensues, as the fearless hounds try to get a mouthhold. The bear often wounds or kills many of his attackers, ripping them open with his fore-claws, or slashing them with his teeth.

Frequently the pack makes it so uncomfortable for the bear that he climbs a tree, whereupon the dogs' voices change pitch, the signal for hunters to move fast, since the first man on the scene usually shoots the bear and gets the hide. Sometimes, the hounds will hold the bear and prevent it from climbing the tree, whereupon a hunter must wade in among the dogs and try to shoot the bear with a pistol. A wounded bear can destroy many valuable hounds.

Bear dogs are usually Plott or redbone hounds or a mixture of these and other breeds. Plott hounds, the preferred breed in our area, were brought to America from Germany by the Plott family, who settled in Maggie, and whose descendants are still there today. Renowned hunters of western mountain lions as well as bears, the brindle-colored Plott hounds have spread across America.

One night in the early fifties, we found a full-grown cow lying dead in a woods road. From the tracks near the scene and marks on the cow, we deduced that a bear had jumped on the cow's back, and after riding her for some two hundred yards, had killed her with a blow to the side of her head. The bear had eaten only the cow's udder. We set traps around her carcass, and that night one trap pulled two toes from the bear's front foot but the bear escaped. For months afterward, we found bear tracks with missing toes around many cattle carcasses. But the bear had learned something from the episode: he never dined twice on a kill but would simply kill a new cow and eat only the udder each time he had a yen for meat.

The following summer my chief guide Glenn Messer was leading a group of horseback riders along the park trail behind a neighbor's pasture. Glenn found a broken rail fence, tracks, and indications that something had been dragged into the park. He followed the trail about twenty yards into the brush where he found the carcass of a fine young heifer covered with broken limbs and leaves.

We set some traps around the carcass. Checking the traps next morning, Glenn found the telltale tracks with two toes missing from one foot, but as usual the trap-wise bear

One of the first sheep-killing bears trapped being skinned, with young Juju Alexander an interested onlooker.

had steered clear of the kill. On his way back to the ranch, Glenn found the carcass of one of my steers, killed the night before.

Realizing that we were contending with a bear impossible to trap, we called our bear-hunting neighbors for help. I couldn't go on this hunt because of a bad foot, but some fifteen or twenty men and their bear dogs did. They spread out along our side of the park line at strategic crossing places, while the dogs were led to the steer's carcass and immediately set off on the track of the bear. The bear crossed into the park, getting through the line of hunters unscathed, while the dogs followed him in and put him up a tree about a quarter mile inside the preserve. When the men arrived, they shot the bear out of the tree. Sure enough, it was missing two toes on one foot. One of my work horses dragged the animal out, and we brought him to the barn loft, where we skinned him and divided the meat among the hunters, as is the custom. All I kept was the foot with missing toes.

But wanting to settle my legal dilemma once and for all, I called park authorities and told them what had happened. Much distressed, they asked me to write a report of the episode and suggested that I might wind up in the penitentiary. Early the next morning before I even had time to write the report, three carloads of park rangers drove up to the ranch to investigate. We showed them the carcasses of the dead heifer and steer, the hole through the fence where the bear had dragged the heifer into the park and hidden it, and the spot where the bear had been shot. The following morning more carloads of rangers arrived to see for themselves.

All this occurred in the midst of a busy haying season, and when on the third morning, an FBI man arrived in the field wanting to talk to me, I told him I was in a jam trying to get hay in and would talk when I finished mowing that patch. He waited and, when I finally joined him, told me he had been sent by the Department of the Interior to investigate.

I took him to my office and explained the trouble I had been having with bears for many years. He expressed sympathy, but three or four days later, I got a telephone call from the federal marshal asking that I come to Waynesville to accept a warrant. I told him I was too busy and that if he wanted to serve me, he could come out to the ranch.

The next day he showed up with a warrant for my arrest, specifying that I was accused of conspiring to molest a bear, bear firearms in the national park, and kill a bear in the

Tom Alexander and Glenn Messer hold up the head of one of the last bears killed on Fie Top.

park. He did not serve it on me, but stuck it back in his pocket, and asked me not to tell anybody that he had been to the ranch. He also asked me to go to the U.S. Commissioner's office next time I was in town and sign a guarantee that I would be available for trial. When I eventually got into town, the commissioner said that the case would be tried in federal court in Asheville in November.

Word got out about the trial, and I was approached by half a dozen lawyers eager to defend me at no cost. I selected a prominent local man, Roy Francis. On the morning of the trial, I picked up four of my neighbors to take to Asheville as witnesses. On the way, we stopped by a Waynesville dairy to have them pack the frozen bear foot in dry ice. Arriving at the courthouse, I looked for a place to stash the foot, not wanting to hold the package in my lap all during the trial. Next to the courtroom, I spotted the open door of the U.S. Marshal's office, and after asking a clerk's permission, slid the package onto a shelf in his coat closet.

My case was called that morning. After all the publicity about the case, the courtroom was packed. Since I had not actually been present at the bear kill, my lawyer didn't call me to the stand, but called instead many of the mountain men who had been there. All testified that I had not taken part in the hunt though none seemed to know just who had killed that bear. When the jury came in with a verdict of not guilty late that afternoon, there was a wild burst of applause from the spectators.

There had been no occasion to introduce the bear foot as evidence, and I simply forgot all about it in the rush to get my neighbors home. But several weeks later, I happened to encounter the federal marshal, who related that he and his staff had had to move out of their offices because of a mysterious smell that had overwhelmed the courthouse and driven most of its inhabitants outdoors before its source was finally uncovered. Good naturedly, he accused me of planting the foot on purpose.

We had gotten rid of the worst killer, but I was disappointed; the trial had set no precedents that would allow livestock raisers to track and eliminate killer bears in the park. Nevertheless, accounts of the trial had spread via the national wire services and a magazine article in the *Saturday Evening Post.* Consequently I got letters from all over the world. One, from a man in Washington state, inquired about how to handle protected elk that were destroying his garden. A man in Wisconsin wondered how to deal with muskrats that made holes in the dirt dams of his fish ponds, draining the water. Two or three people sent checks to help pay my court costs, all of which I returned.

Still, the prominence of the case proved an embarrassment to the authorities. The state finally adopted my old suggestion of hiring an expert tracker to patrol the park border on horseback. The Feds also acknowledged our problem and agreed that we could track and kill a stock-killing bear in the park if a federal ranger were present. When I pointed out that the logistics of bear hunting, including the difficulty of controlling the actions of dogs following a hot trail, often make it impossible to get a warden to the scene in time, one exasperated federal man finally took me aside and suggested that next time I had a bear problem, I "simply go in and kill the bear, and just not tell anybody." Apparently, furthermore, the word must have got to the bears as well. After that time, for whatever reason, we lost relatively few more cattle and killed only a few more bears.

One of the last bear kills occurred in 1953. That summer we had two or three bear traps set at separate cow-kill sites along the face of the bald. Son Tommy was charged with checking traps every morning. He would leave before 6:00 a.m., walk and run the three or four miles of trapline, and usually be back in time for breakfast. One morning he was accompanied by his first cousin, Marshall Roberts, and the pair not only failed to show up for breakfast but still hadn't come in two or three hours later. I finally went out with Glenn Messer to see what had happened. After visiting the various trap sites and doing some tracking and a lot of hollering, we finally located the two boys in the middle of a dense woolly head, or laurel thicket, sitting on a big dead bear.

They said they had found a trap gone from the kill site on the right-hand end of the bald but had trouble tracking which way the trapped bear had gone. It turned out that the trap's "grabs," or grapnel hook, had become caught up in the trap and so not only had it left little sign, but had little effect in slowing the animal. For another thing, the bear had headed across a cleared field, further obscuring its tracks.

The boys had split up and begun circling through the adjacent woods, hoping to strike the track again. Tommy finally did so and followed the trail deep into the laurel slick on Lost Ridge. After crawling through this on hands and knees for several hundred yards, he heard a faint clink, presumably that of a trap chain. Peering over a log, he discerned a low dark shape a few yards away in the gloom and could smell the primeval odor of an over-heated, unhappy bear.

Tommy concluded that the bear was entangled and immobilized and that it was probably safe to move in close enough to make out the animal's head and place a shot. He carried a high-powered rifle but at that point had only one cartridge left: on their way to the bald that morning, the boys had blasted away at some flocks of crows in the meadows, saving only a round apiece for possible business. As Tommy crept in, a sudden commotion and clanking occurred, and he realized that not only was the bear far from immobile but had decided to depart the thicket along the only feasible route he could see, which was the way Tommy was crawling in.

Tommy, on hands and knees, said he commenced backpedaling as fast as he could. He wore glasses for nearsightedness and as often happened, the glasses got torn off in the brush, rendering him practically blind at any distance. At that point he realized he faced a choice: either to try to locate and replace his glasses and shoot the bear or else to keep blundering backward trying to get out of the way of the oncoming animal, which arguably may have been about as interested in avoiding that confrontation as Tommy was. As it happened, Tommy was able to recover his glasses, put them on, and place a fatal shot at close range with his single remaining bullet.

After we had found the boys and their bear, Glenn went back to the ranch to fetch Haynes and a work horse. Meanwhile, we dragged the four-hundred pound bear carcass several hundred yards downhill to a clearing, where we planned to chain him behind the horse and drag him to a road so he could be loaded on a truck. But as the approaching horse picked up the bear scent, he became skittish as some horses do; it was all we could do to hold him still long enough to hook the bear to the singletree. Then as the horse started off he suddenly realized he was being followed by a bear! The last thing we saw was the horse tearing around the bend by the Glenn house, dragging both the bear and little Haynes, still holding the reins and still yelling, "Whoa! Whoa!"

And not long after that episode, Tommy acquired his first set of contact lenses.

✣ ✣ ✣

One frosty November dawn, I was with a party hunting bears near the eastern boundary of the ranch. Several of us were in stands along a ridge and down an old logging road, while others had taken the dogs around the face of Walker Bald, where a bear had been "using," or feeding, all fall.

As the sun rose and the frost disappeared, the morning turned warm and beautiful. Finally around 11:00 a.m., there being no sound from the dogs, we gradually left our respective stands and drifted back to where our cars were parked in a field. Lounging about waiting for the drivers to come in, we occupied ourselves with tales of previous hunts.

One of the party had brought a fruit jar of corn liquor, which we passed around, each taking a healthy swallow. The bottle came around to Bill, who was in the middle of a story and didn't immediately take it from his neighbor. A wayward yellow jacket, feeble and doomed by the approaching winter, came buzzing in, alighted on the mouth of the jar, and crawled inside. Everyone but Bill saw the jacket; nobody said a word. Finally pausing in his story, Bill reached over for the jar and turned it up to his mouth. In went the yellow jacket with the mouthful of whiskey.

Dynamite couldn't have blown Bill up more abruptly. As he sprang to his feet, out came liquor, yellow jacket and all, splattering the laughing group.

"What the hell kind of liquor is that?" he finally managed to ask, rubbing his mouth and tongue.

"What's the matter, Bill, too strong for you?" asked the owner of the bottle, when he finally recovered. It was some time before anybody explained to Bill just what kind of liquor he had imbibed.

✣ ✣ ✣

In wintertime, I often entertain groups of friends who want to try some mountain grouse hunting. After a day of beating across the ridges and through the laurel slicks and shooting up a box or so of shells, maybe with nothing to show for it, we'll gravitate to the big fireplace for a few drinks and some hunting lies. On one such occasion, I happened to glance up at the exterior door just as Judy poked her head in, glanced at the crowd, and gave me a wink. Then she entered, toting an enormous wildcat slung across her shoulder by its hind legs, obviously dead. Its head almost dragged the ground. As the assembled hunters gaped, she headed for the kitchen, saying nary a word.

I collected my wits enough to say, "Got 'im, did you? Whereabouts?"

"Down the road a ways," Judy replied, equally deadpan, disappearing into the kitchen.

It was only much later that she broke down and told the story. Driving up the mountain, she had encountered our neighbor Wesley Gibson with the freshly killed cat. Wesley had simply sold it to her for a dollar after she had expressed interest in the fur.

Miss Judy's wildcat—the one she got "down the road." Cats of this size may be the origin of rumors that "painters" (panthers) still roam the Appalachians.

Since the beginnings of our tenure on Fie Top Mountain, we have always spent our winters here, and in many ways winter is the season we love best. While summertime temperatures at our altitude generally average about twenty degrees cooler than the flat-country regions of North Carolina, during winter we experience alternating systems of frigid weather sweeping down from Canada and warm currents flowing up from the Gulf of Mexico. In wintertime, moreover, the higher elevations of the southern Appalachians experience surprisingly frequent temperature inversions, in which warm air overrides cold, leaving high peaks basking in temperatures that may be twenty degrees warmer than valleys below. The result of these conditions is that in a single twenty-four-hour period our weather can change from a nighttime blizzard, with temperatures reaching twenty below, to balmy afternoons in which we stroll around in shirtsleeves. While our snows can be deep, cold spells rarely last long, so outdoor life is seldom curtailed the way it is further north.

Virtually from the start of the new ranch, I continued to pursue my profession of forestry, often leaving Judy not only to manage the resort but the farming operations as well. In the outdoor activities, she was assisted by a succession of competent foremen, including Larry Caldwell, Rankin Ferguson, and Glenn Messer, but the resort responsibilities she largely handled herself. Later, after our two daughters were married, they and their husbands, Rick Coker, who married young Judy, and Tom Aumen, who married Alice, took over most ranch operations. Son Tommy, who had journalistic ambitions, left after college to work for a succession of newspapers and eventually wound up in New York and Washington, with *Life* and *Fortune* magazines.

After being rejected as too old for military service at the beginning of World War II, I volunteered my services to the federal Office of Price Administration where, as a member of the Washington bureaucracy, I helped set lumber and timber prices for a couple of years while Judy, as usual, took care of things on the mountain.

After the war, I went to work for the Champion Paper and Fiber Company, being responsible for procuring the pulpwood for the big paper mill in Canton twenty miles away. Because the gravel road up the mountain was generally impassable in wintertime, I usually parked the car at the foot of the mountain or halfway up and hiked the rest of the way.

The biggest problem we faced was schooling our children. No school buses could make the trip up the mountain, of course, and Maggie was too far to walk. Tommy, young Judy, and Alice all spent their weekdays either in boarding schools or with a succession of flat-country relatives. When our youngest child, George, came of school age in the late forties, Judy taught him at home, using the correspondence courses of the Calvert System. This teaching took only about two and a half hours of her time a day, but when George reached the sixth grade and was sent off to a real school, he was able to skip a grade. Not long after that, though, he was stricken with acute leukemia. After an emotionally wrenching two years in and out of remission, he died in Boston's Children's Hospital.

Nobody from Cataloochee had ever actually seen a real pair of skis in 1940, when this picture was taken, but foreman Larry Caldwell (left) had gotten the general idea from pictures. He steamed thin boards until they were flexible, bent up the tips, and held them bent with strips of leather. Then he attached harness straps to fasten boards to boots and, with golf clubs for ski poles, he and Miss Judy would plunge down the surrounding slopes pretty much out of control.

Another major wintertime worry was finding ways to keep our loyal ranch hands employed during the long months when the guest ranch was closed. There were always cows to milk, sheep and cattle to feed, and firewood to cut, of course, but these provided work for only one or two hands, usually men of families living on the place. A couple of winters were spent constructing houses on the lots I had sold to Edith Begni and Irma Bain; we also harvested sawtimber and pulpwood, experimented with harvesting wild shrubbery, cleared new ground and planted several steep and eroded areas in pine and spruce seedlings, and even tried raising and milling buckwheat and selling the flour.

Shortly after first arriving on the mountain, though, it had occurred to me that the newfangled sport of skiing, which was spreading throughout New England and the West, might also be feasible here as well. I had long ago indicated the location of a possible ski run on a panoramic photograph of the future ranch layout that I shot from Moody Top in 1937. But a study of weather patterns in succeeding winters showed we would never be able to count on nature to provide enough snow for dependable skiing.

Even so, during the early forties, Judy and the kids and foreman Larry Caldwell constructed their concept of skis, which none had actually seen, from thin strips of scrap lumber, boiling it in water and bending up the tips. They bound these appliances to their feet with old harness straps. Without benefit of lessons and using golf clubs for ski poles, they would plunge off the steepest slopes, without much control but somehow managing to survive without serious injury.

Hugh Morton–Cataloochee Ranch Collection

By 1961, Tom and Judy had learned enough about the newfangled technology of snow making to experiment with skiing in North Carolina. Their first run was installed on the hill behind the ranch house. Along with snow guns and rented air compressors, they installed a rope-tow, provided rental skis, and turned one of the original Verlin cabins into a warming hut.

By the late fifties, I became aware of the recent development of snow-making technology, wherein a mixture of water and compressed air, shot from nozzles when the air temperature is near the freezing point, emerges in the form of snow. In 1959 and 1960, Judy and I made a couple of trips to various New England ski resorts that employed snow guns to see whether they would work in the southern Appalachians. The New England resort managers were extraordinarily helpful with advice about the new technology and the ski business in general. Subsequently, I got more advice from consultants about what it would take to start a ski business in the South.

We decided to start small. In the late fall of 1961, we installed a rope tow on the little hill behind the ranch house, a pump beside the fishpond, leased four diesel air compressors, bought some rental skis and boots, converted the cow barn into a ski lodge, and hired an Austrian instructor. When the first good cold snap arrived in December, we spent some frustrating nights learning the tricky art of snow making but finally managed to open the first ski area south of Virginia shortly before Christmas.

Southerners took to it instantly, literally falling all over themselves to experience that wonderful winter sport. We even showed a little profit that first winter. After a couple of years, we graded a better slope and installed a T-bar lift. Finally, in 1968, we bit the bullet and moved the whole operation across the ridge, built a new lodge, installed a lot more snow-making capacity and an expensive chair lift up the north face of 5,400-foot Moody Top. I appointed sons-in-law Rick Coker and Tom Aumen as general manager and mountain manager respectively, while daughters Judy and Alice ran the rental and ski-equipment shops.

While the ski business has always been touch and go from the standpoint of profitability, what has been most important to me is that I was finally able to provide legitimate wintertime employment to family and local hands as well as an economic boost to a local community that had hitherto depended solely upon summertime tourism for income. Since then, nearly a dozen other ski areas have opened in North Carolina and the nearby mountains of North Georgia and East Tennessee, attracting skiers from as far away as Florida.

Alice Alexander Aumen

Miss Judy, Juju, Alice, Tom, Foxie, and Tommy on the ranch house porch in 1939.

Portrait of a Father

BY JUDY ALEXANDER COKER

What was it like growing up here? That question is asked many, many times and my pat answer is that it had its good points and its bad. In reality, I can't remember much of the bad.

The center of my childhood world was "my mountain." Wherever Dad went I tried to be there, and this led me to the hay fields, the woods, the potato patch, among the work hands, and, of course, to the barn. But I was not there as an idle bystander. Dad believed in productivity, and everything, including me, had to produce something worthwhile. My interests lay in the outdoors, so I learned about haying, forestry, gardening, mountain people, and horses. Horses became my primary concern, and I naturally wound up in the barn, leading rides, taking pack trips, and eventually being in charge of the entire riding program. At age thirteen, I was the youngest licensed guide in the National Park System. My teenage years were spent on the trail on my mare Dixie, enjoying a life I assumed to be no different from that of any young girl.

While Mother and Dad recognized my love for the mountains, they believed it imperative that we have first-class educations and learn about the "other world." Being so remote and serviced by no bus system, elementary schooling was a problem they had to deal with. I attended grades one through eight at various local boarding schools, being driven in on Monday mornings and returning on Friday evenings for cherished weekends. For the high-school years I was sent to St. Catherine's in Richmond, Virginia. This was true culture shock: my homesickness literally made me ill, and Mother and Headmistress Turner conducted many a phone call over what to do about me. But I adjusted and survived and will never forget nor regret the four years at St. Cat's.

I went on to Duke University and a major in political science. After graduation I got my first job in the outside world as the personal secretary of Senator Kelly Bennett in our North Carolina General Assembly. In 1957 I married Rick Coker of Hartsville, South Carolina. In a few years Mr. Tom and Miss Judy had three grandchildren: Richard Coker, Jr., Judy Barksdale Coker, and Mary Morton Coker. Each of these kids now works or plays on the mountain year round.

Mr. Tom jumping Prince in about 1942.

I have asked some of Dad's old friends and former Cataloochee guests to share their recollections and thoughts about him. Dr. Charlotte Maguire and her husband, Raymer, had a long friendship with Mr. Tom and Miss Judy. She recalls some of her memories:

> Beginning in the late forties, my husband and I set out to find a place in the North Carolina mountains where we could relax from our respective practices of law and medicine. While attending a pediatric-obstetric seminar, Dr. Alec White of Asheville suggested that he take us to the finest place in the Smokies and introduce us to Tom and Judy Alexander, the greatest of hosts anywhere, who owned and operated Cataloochee Ranch. We went with Dr. White and stayed for a few days, and from then until now we have returned every summer and sometimes in winter for rest and relaxation and for the hospitality of Tom and Judy and their family. Judy was the best provider of wonderful food three times a day as well as for birthdays and holiday celebrations and pack trips. We liked it so much that we bought a place on the mountaintop and built a house.

Slim and modest, "Turkey" George Palmer acquired his nickname as a youngster after devising a clever technique for trapping wild turkeys. He would dig a kind of tunnel leading into a covered log pen and bait it with corn. The turkeys would follow the bait into the pen and then be afraid to return the way they had entered. Palmer's problem arose when it came to removing the live turkeys from the pen; the big, angry birds would sometimes beat him nearly senseless, to the vast amusement of his neighbors.

Courtesy of Great Smoky Mountains National Park

We had great fun and much excitement on those early pack trips with Tom and wranglers like Todd Henry and Sam Woody. They always told tall tales and pulled many a prank on the riders, especially after we pitched camp in the evenings. On one trip, we carried along the two-way radios that the ranch had just acquired so that it could stay in touch with riders far off in the wilderness. We camped one night in Cataloochee Valley at the Turkey George homeplace, one of the few camps in the park equipped with that "modern convenience," a two-hole outhouse.

Tom and his wranglers hid one of their new-fangled radios down in one of the holes in the outhouse and waited. One rider, a prim and proper lady, found it necessary to visit the outhouse shortly after dinner. She went in, closed the door, and after a half-minute or so a voice boomed out of the hole, "Hey, lady, would you mind moving over? We're still trying to paint down here." Out she came, screaming, clothes half on and half off.

Judy and Tom became special friends and visited with us in Orlando and Tallahassee from time to time. After Raymer died, Mr. Tom and Miss Judy played an important part in my son Thomas' life. He worked on the ranch a couple of summers and was even married in the ranch house. After Tom's stroke, Thomas would come up from Athens where he was in college and take Tom riding or drive him to some favorite place.

✣ ✣ ✣

Dad's real profession was forestry, and we have an interesting account of his business dealings from George H. V. Cecil, a grandson of George Vanderbilt who founded the 125,000-acre Biltmore Estate and constructed the palatial 255-room Biltmore House in the late nineteenth century:

Consulting forester Tom Alexander, making notes in the field. His profession often kept him traveling or commuting to a daily job at the Champion mill at Canton. This picture won Alice Alexander first prize in the "Best Looking Father" contest at St. Catherine's School in Richmond.

The earliest recollection that I have of Tom Alexander was during the summer of 1947. Reuben Robertson of Champion Paper Company had approached Judge Adams, then president of The Biltmore Company, to explore the possibility of a timber-cutting program on the Biltmore Estate.

In the late nineteenth century, the estate had hired the famed conservationist Gifford Pinchot to supervise planting of the first large-scale managed tree plantations in the United States. Later, under German forestry expert Dr. Carl Schenck, these plantations formed the heart of the famed Biltmore Forestry School, now known as the "Cradle of Forestry." Those half-century-old plantations were what we initially contemplated thinning. Besides profits from pulpwood sales, the program would benefit Biltmore inasmuch as the forests were ready to be thinned. Moreover, if properly planned and executed, the program might demonstrate that it was possible for paper companies and landowners to work together for the benefit of both and not at cross purposes, as had so often been the case in the past.

To implement this program, Champion sent their chief forester, Tom Alexander, to work out the details with Chauncy D. Beadle, then superintendent of the estate. Mr. Beadle, who had come to Biltmore in 1890 as an assistant to Frederick Law Olmstead and had stayed on at my grandfather's specific request, had quite obviously a very keen sense of what he thought was right and how he wanted "his forests" handled. Had it not been for Tom Alexander, it is doubtful that the program would have been as successful as it turned out to be.

A firm friendship and respect for each other developed between Alexander and Beadle. Tom was honest enough not to try to seek a short-term gain to the detriment of the forests as a whole and realized that here was a perfect opportunity to practice the best of forestry. Beadle recognized this, had complete faith in Tom's honesty and integrity, and basically left the planning of the program to him. The contract had the provision in it, quite naturally, that Biltmore could employ its own forestry consultant to check on Champion's program, but in actual fact this clause was rarely invoked, such was the trust that Beadle had in Tom's integrity.

The contract that evolved was a masterpiece of checks and balances. Quite frankly, though, it was probably more of an evolution than a preplanned program. In essence the program worked like this—Tom would plan for the overall area which was to be cut or thinned, depending on the terrain or the predominant species of tree. Once the loggers had cut through an area, the sawlogs were scaled, hauled, and sold to a lumber company. The pulpwood would go to Champion and be paid for on a per-cord (later per-pound) basis. All scaling was done by Champion, which had no interest in the sawlogs. Contrary to then-current practice, the logs were scaled using the International Rule as opposed to the Scribner or Decimal. This was agreed to at the outset as being the fairest scale for all parties. It caused quite a bit of confusion, but was rigidly adhered to over the thirty-plus years that the agreement was in force.

Parenthetically, as I recall, the only time the checks and balances got out of "sync" was in the mid-seventies when Champion had a new scaler, unfamiliar with the International Rule, and the purchaser of the logs complained bitterly that he was not getting the yield he anticipated. The problem was quickly solved to everyone's satisfaction.

N.C. State News Bureau–Cataloochee Ranch Collection

Tom also had to act as peacemaker between Judge Adams and Reuben Robertson. The judge, as he was always known, had both a Lincoln Continental car and a very fiery, if brief, temper. One day the judge was driving around through the woods (where, quite frankly, he should not have been due to the logging), and got himself inextricably stuck in the mud about two miles from the nearest ranger house. The explosion that followed had to be heard to be believed, and before the day was over, a letter was on its way to Champion, the gist of which was to get their blankety-blank loggers off the property and never to return! Thanks to some very diplomatic work by Tom Alexander, peace was eventually restored and logging resumed.

It was with this background that I, personally, came to know and respect Tom Alexander. We would tramp through the woods together, and he would explain to me how the program worked and how and why he planned treating the area under study—usually thirty to fifty acres at a time. This friendship matured to

Sunday night cookouts like the one at left have always been a Cataloochee Ranch tradition. Right, Juju and Alice Alexander with an assortment of the antique ironware often used for outdoor cooking at the ranch. Mr. Tom always presided as outdoor chef until the early fifties when Norman Wingate (below) sometimes took over. Behind Norman is Jim Lamb who, along with his sister Marcia Zarger, has been coming to the ranch since the first Cataloochee opened.

Alfred A. DeLardi–Cataloochee Ranch Collection

the point that Tom and Judy invited me up to the ranch for the day. We rode horseback, partook of some of the refreshments he used to keep under the bed in the cabin they occupied during the summer when the ranch was open to guests, and as I recall, enjoyed a delicious steak cookout under the stars. We at Biltmore used to swap bull calves for riding privileges at the ranch.

One ride that particularly stands out in my memory was a homecoming at the old Cataloochee Valley church. We all left the ranch and rode out over Polls Gap. Once down in the valley, Tom and I left the group to look at and roughly measure some enormous poplar trees in a grove uphill from the trail. We pulled some vines and, by stepping off their lengths, used them as crude tape measures to determine the circumference of the trees. It became apparent that the trees might be of record size. I understand that Tom went back later, measured them more accurately, and that they subsequently found their way into the list of record-sized trees. After this, we rode on down to the church where we met Judy, who had driven a station wagon over there loaded with food for what by then were very hungry riders!

During the early years of my marriage, due to my responsibilities of raising a family, I did not see Tom as much as earlier. This all changed, though, in the early sixties when the first ski slope was opened. Nancy and I used to bring our children up what by then had become a vastly improved road to enjoy a day's skiing. Many a time we would come on Saturday and Sunday as well as on any holiday that fell during the week. Our paths crossed frequently then, and I can remember many happy hours visiting with him.

One of the greatest tragedies was to watch Tom's health and eyesight fail. He had been such an outstandingly striking person with his white hair and ruddy complexion. I never heard anyone hold him in anything but the highest esteem, and I even have a feeling that the Park Service, which had to take an adversarial position on occasions, respected him for standing up for what he felt was right.

✣ ✣ ✣

Jim and Marcia Lamb Zarger have returned almost every summer since 1935, and Marcia recounts some of her memories:

In June 1935, I was met in Waynesville by Tom's brother, Miller Alexander, who said he was sorry for being delayed but that he had run off the road a little and been forced to return to the ranch for another car. At Cove Creek Gap we stopped for the view—my first glimpse of Cataloochee Valley. I noticed that Miller's hand shook as he lit a cigarette, but I was not prepared for our next stop—by the roadside on the way down the mountain. Over the edge, some forty feet below us, I had my first glimpse of Tom Alexander by the wrecked car. He waved and then hiked up the slope to greet us. His appearance was striking—white hair, sky-blue eyes that smiled, an outdoors tan—and he had a hearty handshake

Jack Szold–Cataloochee Ranch Collection

Pack trippers in the Smokies have to be prepared for Smoky Mountain weather. Led by wrangler and trail cook Sam Woody, a party of riders pauses to rest horses during a foggy day in the national park in 1953.

that made you like him immediately. As a host he was genuine, took one at face value, and always displayed his great sense of humor.

A fascinating aspect of a stay at the old ranch was the rough and ready lifestyle of the Cataloochee Valley. One day, Larry Caldwell and I returned a wagon to the farm of Tommy Caldwell, who also functioned as the valley's unlicensed dentist. It happened to be tooth-pulling day, and the children were all lined up. Tommy would ask one to open his or her mouth and ask, "Which one?" and the child would point. Tommy then put in his pliers and out came the tooth—hopefully the right one. Then Tommy would say, "Go rinse your mouth with salt water. Next!"

In 1939 Tom organized a three-day pack trip to Ledge Bald. Six of us, Bonnie and Arnold Griswold from Louisville, Irma Bain of Lexington, Tom, Larry, and I, set out. The first afternoon it poured and it was wet setting up camp, but we had a nice roaring campfire. The next morning after breakfast we started to saddle up. The saddles were wet, the bushes were dripping—no real fun to ride, but we were out for a three-day pack trip and a three-day pack trip it would be. The men fished and we sat around the fire and spun yarns. All socks, spare and regular, were wet, so we impaled them on sticks around the fire and picked out the driest whenever we left the fire. We rode into the ranch declaring it the best rainy-day pack trip ever.

I remember Tom calling square dances up in the barn with Cal Messer and his fiddle, Mark Hannah and his autoharp, and Rufe Sutton and Aletha Wood dancing. Singing and playing a variety of tunes, some composed by Cal—"I Had One Old Hat," "On a Log in a Hole at the Bottom of the Sea," "Lamplighting Time in the Valley," "Going Down the Road Feeling Bad," and, of course, Tom's favorite in the later hours of evening, "Will There Be Any Stars in My Crown?"

At breakfast one morning in 1941, Tom asked us if we would like to visit a whiskey still. Apparently he had stumbled upon one on ranch property the day before and had worked out a deal with the operators. All the guests said yes and off we rode on horseback. We rode about half an hour, circling around through the woods, on and off the trails, and finally dismounted and tied up our horses. Then we were blindfolded and led through the brush to the still. There was the fire, the vat, the coils, the cooling water, and the white lightning in half-gallon jars. We all had a taste, had our picture taken for the record, were blindfolded again, and returned home by another circuitous route.

✣ ✣ ✣

Bob Russell plays mountain music with a small band he has gathered together. He remembers as a boy of twelve being invited to sit in with the group that played for ranch square dances. At the end of the evening Mr. Tom stuck a dollar in the young boy's shirt pocket. "I was so proud, I left the mountain feeling like a king," says Bob.

✣ ✣ ✣

Susie and Bob Cowden first came to Cataloochee in October of 1950. Their first impression of Miss Judy and Mr. Tom was so positive that they quite forgot the dreadful road they had just climbed and settled in for their two-week stay. As many others have done, they eventually acquired a mountaintop home and have been part-time residents ever since. Susie gives us this account:

That first night at dinner we were seated next to Mr. Tom. He was the handsome, sun-tanned, white-haired, blue-eyed southern gentleman at the head of the table. He and Bob took to each other immediately. There began a friendship which lasted the rest of their lives. For many years, Tom and Bob spent time sitting in our front yard, looking out at the beautiful mountains and discussing what was on their minds. Sometimes they had serious concerns. Sometimes they shared problems. Sometimes what they talked about produced hearty laughter.

I recall that on that first trip when we were staying at the ranch, we had to try to sort out which parts of the mountain stories were true and which were tall tales. We were determined to resist being fooled by the tall tales. One Sunday

Aycock Brown—Cataloochee Ranch Collection

Picking and fiddling are mere sidelines for most musicians who have ever played at Cataloochee dances. Cal Messer, the fiddle player above, was a tobacco farmer and Rufe Buff, the banjo player, was among other things, a logger. But when they truly fell into rapport in some old tune, they attained a spontaneous, driving gusto that professionals seldom manage. Originally, square dances were held almost weekly in the ranch living room. Later they were moved to the barn loft.

J. Baylor Roberts—Cataloochee Ranch Collection

morning Tom asked if we would like to ride horseback with him up to Hemphill Bald to salt the cattle. When we got to the open grassy place, Tom said it was time to call in the cattle and began yelling, "Woo-woo, woo-woo, woo-woo." Bob and I looked at each other; we were not going to be fooled into believing cows could be summoned the way a dog could. We were not going to fall for that one. So we rode past Tom and on up to the top of the mountain. When we glanced back, there was Tom surrounded by cattle licking the salt he had brought for them. Sheepishly we turned around and came back down. We had a lot to learn about the way of the mountains.

On one of our visits to the ranch, Tom and Judy asked if we would like to join them on a pack trip on horseback along the Appalachian Trail. We jumped at the chance and soon found ourselves riding along what seemed to be the top of the world. We were looking down at Tennessee on one side and at North Carolina on the other. In some places the land dropped away, and this was the very place the horses insisted on walking on the very edge of the trail. Tom told us to rein our horses toward the bank, away from the edge. That worked fine until we came to a section of the trail where the land dropped away on *both* sides. In moments like that I sort of leaned toward the North Carolina side. After getting to know Tom and Judy, I decided I'd rather die in North Carolina, the place they were teaching us to love as much as they did.

Our friendship started in 1950. Now it is forty-four years later. Both Tom and Bob have died. But the memories of that friendship and the influence that it had are as clear and strong as ever.

✣ ✣ ✣

Cookie Wood worked on the mountain from the time he was old enough to pick up 'taters for ten cents an hour until he found full-time work at another occupation. He remembers always feeling at ease around Mr. Tom: "'While it's rainin',' Tom would say, 'let's me and you take that old snub-nosed jeep around and see if we can scare up any grouse-eatin' foxes.'"

Cookie tells of one occasion involving a farm hand with the habit of taking ranch pitchforks home. "Tom finally got tired of buying pitchforks, and one day he just said to the man, 'When you come to work in the morning, how about bringin' in all those forks you have down there. We'll need 'em all, dry as it's gettin'.'"

✣ ✣ ✣

Babs and Arthur Chenoweth started their summer visits to Cataloochee in the early fifties, and, like many other former guests, eventually acquired their own place on the mountain. Arthur recalls:

> Tom had dignity, but part of that dignity was his earthiness. On one lazy day, Tom and I were sitting on the porch of Laurel Cabin sippin' a bit of bourbon,

Many former youngsters, now grownup friends of Cataloochee, remember Mr. Tom's evening story telling and his readings from such classics as the Uncle Remus tales high points of their stay. This picture was taken in 1953.

talking about nothing in particular, just enjoying the day and each other's company. From the corner of my eye I noticed something white and mobile and turned my head to see a white groundhog coming out from under a log. Not really believing my eyes, I asked Tom, "Did you see that white groundhog?"

Tom replied, "Thank God! I thought I'd had too much and was seein' things."

✣ ✣ ✣

Bettina and George Shackelford are former guests and now part-time residents at Cataloochee. Writes Bettina:

> Mr. Tom had presence. It showed in his stance, in his snapping eyes, and in his wonderful voice. He was quite able to take on the Feds, the Park Service, and anyone else in the matter of a marauding bear, perhaps the very bear that was dished up for dinner the very first night we spent at Cataloochee.
>
> But he had humor and gentleness and it shows in a sharp and much-loved picture of our two young boys, Peter and Ruffin, who were perhaps four and six, sitting on the arms of Mr. Tom's chair, listening raptly to readings of the Jack Tales and Uncle Remus. Tom was reading to the children, but every adult in that big room was listening.

Peter can tell you that when Mr. Tom spoke, a boy really listened. Once, Peter had the misfortune to leave open a gate that should have been shut. After Mr. Tom told him about gates, he never forgot the rule of leaving gates the way you find them.

Miss Judy was as much a part of the ranch as Tom, and I don't think that there was anything she couldn't handle nor any person she didn't like or at least find something good about. She took care of Peter's horse when the front hoof was badly cut and Peter was so worried. She concocted some revolting brew for me when I had a tummy ache and she stayed to make sure I drank it. Either it worked or I was afraid she'd stir up another dose, because I got well.

Miss Judy always had some hot sauce for us because we came from Texas, but her very best dish was the macaroni and cheese that went along with Sunday night suppers. She usually sprinkled paprika over the top for La Presentación, but on one particular Sunday her hand went astray and the casserole was liberally covered with cayenne. It looked great, and we liked it; I'm not sure about the other guests.

The long dining-room table with Miss Judy at one end and Mr. Tom at the other was a reward for the day of riding and hiking. When you first came, your name would be on a clothespin somewhere in the middle. Then you moved up to sit near Miss Judy. If she found you interesting you were moved next to Mr. Tom. That was an honor. But always Miss Judy was there to make sure you had had a good day.

After a little time with Miss Judy and Mr. Tom, you knew that you had been given a gift—a gift of sharing the beauty of Western North Carolina, of being with two people who accepted you just the way you were, who expected and got the best from most people, and who, I think, always gave the best back to all.

✣ ✣ ✣

The central philosophy in our upbringing was that we could do anything we set our minds to. Never do I remember hearing the words "You're too young," "Watch out! You'll get hurt!" or "I don't think you can." If we wanted something, get it. If we wanted to do something, do it. But always be fair and honest with all.

As I grow older, I find myself more and more impressed with the farsightedness and depth of Dad's planning and his ability to get things done, physically and politically. He admired the independent and resourceful mountain people, who by and large lived by the philosophy of do it yourself or do without. His dignity and sense of self-worth enabled him to walk and be comfortable in any circle, and he tried to teach us that.

Alice Alexander Aumen

Cataloochee Ranch, with the Soco Gap beyond, viewed from Moody Top. This view, the same as shown in the endpapers, shows how the ranch complex developed compared to Mr. Tom's original concepts.

Alice Alexander Aumen

Looking southeast from Hemphill Bald toward Moody Top and Ned's Licks. The road connects the ranch area at right with the ski area at left.

The view westward toward Soco Gap from Moody Top. When natural snow is deep enough, Moody Top offers skiers a close approximation of western-style open-bowl skiing.

Alice Alexander Aumen

Epilogue

My father died of the consequences of a stroke in 1972 at the age of 72, not long after completing the chapter on Fie Top. As of this writing, Miss Judy is in her late eighties and still going strong, though she no longer plays an active role at Cataloochee Ranch and Ski Area, which continue largely under the operation of Alexander children, grandchildren, and spouses. It is clear that the Alexander strain of mountain fever is virulent and genetically transmitted. As for me, I have given up career pursuits in such garden spots as New York City and Washington, D.C., and, as most mountain boys eventually do, come home to the hills. My wife Jane, originally from Utica and Long Island, New York, has adapted well to log-cabin life and mountain ways.

As a memorial to Mr. Tom and Miss Judy, we descendants have deeded Hemphill Bald and other portions of ranch forest and meadowland to a permanent conservation trust, guaranteeing that it will never be developed but will remain in its pristine state for the benefit of the public. The terms of the trust are administered by the Southern Appalachian Highlands Conservancy, an organization dedicated to preserving the scenic, recreational, and wildlife resources of the southern Appalachians in the face of intense development pressures. It is an organization that our father would have applauded, both for its preservation instincts and for acknowledging that there are human as well as ecological values to be preserved.

Tom Alexander, Jr.

Jane Alexander

Hemphill Bald, looking south toward Utah Mountain, now permanently preserved from development.

Notes on the Photographs

All photographs are the property of Cataloochee Ranch unless otherwise credited. Many have been published previously in feature stories on Tom Alexander or the resort, and when the photographer could be determined with reasonable certainty, credit has been given. Similarly, members of the Alexander family or friends whose photographs were loaned to the publisher for inclusion in this book are gratefully acknowledged.

Bright Mountain Books, Inc., is indebted to the archivists and library staff members who aided in locating pictures taken in Western North Carolina during the early part of the century. Often little more than the name written on a photograph is known about the photographers whose pictures were chosen, but the following facts are known:

> Frank W. Bicknell, originally from Iowa, lived for a time in Upper Creek Township, Burke County, North Carolina. He worked as a photographer and journalist in Western North Carolina from about 1905 to 1920.
>
> Masahara Iizuka (George Masa) was born in Japan in 1881 and studied mining engineering in Tokyo at Meige University and later at the University of California. Coming to Asheville in 1915, he soon became involved in news photography while spending as much of his time as possible in the Great Smokies measuring trails and taking photographs. His expertise was drawn upon by the Nomenclature Committee during the formation of the Great Smoky Mountains National Park. With Asheville newspaperman George W. McCoy, Masa coauthored the *Guide to the Great Smoky Mountains National Park* in the 1930s. He died in 1933 in Asheville.
>
> Margaret Warner Morley was born in Montrose, Iowa, February 17, 1858. As a biology teacher in Brooklyn, New York, she wrote prolifically on many subjects, authoring textbooks on nature studies and conservation. Miss Morley spent several summers in Tryon, North Carolina, photographing and writing about the area, resulting in the publication of *Little Mitchell; the Story of a Mountain Squirrel* in 1904 and *The Carolina Mountains* in 1913. She died in Washington, D.C., December 12, 1923.
>
> Herbert W. Pelton, a native of Haywood County, North Carolina, is renowned for the panoramic views of urban life taken with a Cirkut camera. Pelton shared the Plateau Studio in Asheville with George Masa before moving to Washington, D.C., in 1924, where he continued his photographic career. The Library of Congress Collection owns many of Pelton's panoramas.
>
> R. Henry Scadin lived in Michigan from his birth in 1861 until the 1890s when he and his wife and daughter moved to Sapphire, North Carolina. Many of the scenic views Scadin searched out in the Highlands-Brevard-Tryon area exist on fragile glass plates, a form of negative which is becoming increasingly rare. Scadin's later years were spent doing commercial photography and pursuing a life-long interest in horticulture. He died in 1923.

We wish to thank the following for their kind permission to reproduce photographs:

George Ellison (pages vi and 56)

North Carolina Division of Archives and History (pages xi, 7, 10, 20, 34, 62, and 133)

Ramsey Library, University of North Carolina–Asheville (pages xiii, 2, 5, 13, 22, 25, 36, 37, 40, and 72)

Barbara Ambler Thorne (pages 49, 56, and 57)

Great Smoky Mountains National Park, Gatlinburg, Tennessee (pages 72, 83, 87, 89, and 147)

Hunter Library, Western Carolina University, Cullowhee, North Carolina (pages 130 and 133)

1/10th of property shows in this pict
moody top, and sheep pasture
Plant Balsams
Lake to be
Oak grove
now
ski run